Studies in Systems, Decision and Control

Volume 108

Series editor

Janusz Kacprzyk, Polish Academy of Sciences, Warsaw, Poland
e-mail: kacprzyk@ibspan.waw.pl

About this Series

The series "Studies in Systems, Decision and Control" (SSDC) covers both new developments and advances, as well as the state of the art, in the various areas of broadly perceived systems, decision making and control- quickly, up to date and with a high quality. The intent is to cover the theory, applications, and perspectives on the state of the art and future developments relevant to systems, decision making, control, complex processes and related areas, as embedded in the fields of engineering, computer science, physics, economics, social and life sciences, as well as the paradigms and methodologies behind them. The series contains monographs, textbooks, lecture notes and edited volumes in systems, decision making and control spanning the areas of Cyber-Physical Systems, Autonomous Systems, Sensor Networks, Control Systems, Energy Systems, Automotive Systems, Biological Systems, Vehicular Networking and Connected Vehicles, Aerospace Systems, Automation, Manufacturing, Smart Grids, Nonlinear Systems, Power Systems, Robotics, Social Systems, Economic Systems and other. Of particular value to both the contributors and the readership are the short publication timeframe and the world-wide distribution and exposure which enable both a wide and rapid dissemination of research output.

More information about this series at http://www.springer.com/series/13304

Ewa Ratajczak-Ropel · Aleksander Skakovski

Population-Based Approaches to the Resource-Constrained and Discrete-Continuous Scheduling

Ewa Ratajczak-Ropel
Department of Information Systems
Gdynia Maritime University
Gdynia
Poland

Aleksander Skakovski
Department of Navigation
Gdynia Maritime University
Gdynia
Poland

ISSN 2198-4182 ISSN 2198-4190 (electronic)
Studies in Systems, Decision and Control
ISBN 978-3-319-87422-7 ISBN 978-3-319-62893-6 (eBook)
DOI 10.1007/978-3-319-62893-6

Printed on acid-free paper

This Springer imprint is published by Springer Nature
The registered company is Springer International Publishing AG
The registered company address is: Gewerbestrasse 11, 6330 Cham, Switzerland

Foreword

Population-based approaches have proven to be an effective and practical tool for solving wide variety of the difficult optimization problems. This is particularly true with respect to combinatorial optimization where analytical methods have only limited application possibilities. The presented book tackles two of the most difficult and computationally intractable classes of problems. The first is the discrete resource-constrained scheduling, and the second, the discrete-continuous scheduling. Problems belonging to the first class are investigated in the first part of the book by Dr. Ewa Ratajczak-Ropel. Problems belonging to the second of the above classes are dealt with by Dr. Aleksander Skakovski. Both authors have been working on the respective problems during the last decade gaining scientific recognition through publications and active participation in the international scientific conferences. Both authors base their results on applying population-based methods. Dr. E. Ratajczak-Ropel explores multiple-agent and A-Team concepts, while Dr. A. Skakovski focuses on evolutionary algorithms with particular attention to population learning paradigm.

Dr. Ewa Ratajczak-Ropel, in her part of the book, discusses techniques for agent-based optimization, presents the resource-constrained project scheduling models, and briefly reviews various algorithms and approaches proposed for solving them. The core part of her results includes designing and validating several multi-agent systems under the team of agents (A-Team) umbrella. The proposed A-Teams have proven to be an effective tool for solving the resource-constrained project scheduling problems.

Dr. Aleksander Skakovski, in his part of the book, defines the discrete-continuous scheduling problem and discusses its properties. In an extensive state-of-the-art review, several approaches to solving instances of the problem at hand proposed so far in the literature are presented including the heuristic and metaheuristic algorithms. Main results of the author include the proposed island-based evolutionary algorithm, population learning algorithm, cross-entropy-based population learning algorithm, population learning with differential evolution, and island-based differential evolution algorithm. The performance of the proposed algorithms is evaluated experimentally.

Both parts of the book together offer a valuable insight into the possibility of implementing modern techniques and tools with a view to obtain good quality solutions to practical and, at the same time, computationally difficult problems. The book is, in my opinion, an important source of knowledge to practitioners dealing with the real-life scheduling problems in industry, management, and administration.

February 2017

Prof. Dr. Piotr Jędrzejowicz
Gdynia Maritime University
Gdynia, Poland

The original version of the book was revised:
For detailed information please see Erratum.
The erratum to the book is available at
10.1007/978-3-319-62893-6_13

Acknowledgements

We would like to gratefully and sincerely thank our mentor and advisor Prof. Dr. Piotr Jędrzejowicz, Head of the Department of Information Systems at Gdynia Maritime University, for the idea of this book, motivation, continuous support, immense knowledge, patience, and time offered to us. We also thank him for the foreword to this book.

Our sincere thanks to our colleagues from the Department of Information Systems and Department of Navigation for their mental and technical support.

We would like to give our special thanks to our families for their love, help, spiritual support, and understanding.

February 2017

Ewa Ratajczak-Ropel and Aleksander Skakovski
Gdynia Maritime University
Gdynia, Poland

Contents

Part I Agent-Based Approach to the Single and Multi-mode Resource-Constrained Project Scheduling

1 Introduction 3
References 5

2 Agent-Based Optimization 7
2.1 Basics of the Agent-Based Approaches 7
2.2 Agents-Based Approaches to Optimization 10
2.2.1 A-Team Concept 12
2.2.2 A-Team Implementation - JABAT 14
2.3 Agents-Based Approaches to Project Scheduling 19
References 19

3 Project Scheduling Models 25
3.1 Historical Review 25
3.2 Basic Models and Classifications Review 26
3.3 Generalizations and Special Cases of the RCPSP 29
3.4 Objective Functions 30
References 30

4 Resource-Constrained Project Scheduling 33
4.1 Problem Formulation 33
4.2 State of the Art Review 35
4.3 Agent-Based Approaches to Solving RCPSP 37
4.4 A-Teams Solving the RCPSP 39
4.4.1 Single A-Teams with the Static Cooperation Strategies 40
4.4.2 Algorithms Used in the Further A-Team Approaches 44
4.4.3 Randomized Team of A-Teams with Static Cooperation Strategy 49

4.4.4 A-Team with the Dynamic Cooperation Strategy with Reinforcement Learning 50
4.4.5 A-Team with the Dynamic Strategy Based on Population Learning 54
4.4.6 A-Team with Dynamic Cooperation Strategy Based on Integration 56
4.4.7 Concluding Remarks 59
References 62

5 Multi-mode Resource-Constrained Project Scheduling 69
5.1 Problem Formulation 69
5.2 State of the Art Review 71
5.3 Agent-Based Approaches to MRCPSP 73
5.4 A-Teams Solving the MRCPSP 76
5.4.1 Single A-Teams with the Static Cooperation Strategies 76
5.4.2 Algorithms Used in the Further A-Team Approaches 79
5.4.3 A-Team with Dynamic Cooperation Strategy with Reinforcement Learning 85
5.4.4 A-Team with Dynamic Cooperation Strategy Based on Population Learning 87
5.4.5 A-Team with Dynamic Cooperation Strategy Based on Integration 88
5.4.6 Concluding Remarks 89
References 94

6 Conclusions 99

Part II Population-Based Approaches to the Discrete-Continuous Scheduling

7 Introduction 103

8 Discrete-Continuous Scheduling Problem 107
8.1 General Resource-Constrained Scheduling Problem 107
8.2 Practical Applications of the DCSP 108
8.3 Notation 108
8.4 Task Models 110
8.4.1 Processing Time Versus Resource-Amount Model 110
8.4.2 Processing Rate Versus Resource-Amount Model 110
8.5 Problem Formulation 111
8.6 Variants of the DCSP 112
8.7 General Approach to Solving the DCSP 113

8.8 Main Properties of Optimal Schedules 114
8.8.1 Convex Functions $f_i \leq c_i \cdot u_i$, $c_i = f_i(1)$ 114
8.8.2 Concave Functions f_i and $n \leq m$ 115
8.8.3 Concave Functions f_i and $n > m$ 115
8.9 Minimization of the Maximum Lateness $L_{\max}$ 121
8.10 Minimization of Mean Flow Time F 121
References 123

9 State-of-the-Art Review 125
9.1 Theoretical Research on the DCSP 125
9.1.1 Another Formulation of the DCSP 125
9.1.2 The New Approach to Optimal Resource Allocation 126
9.1.3 New Properties of the Discrete Part of the DCSP 128
9.2 Discretisation of the DCSP 129
9.2.1 Discretisation of the Continuous Resource 130
9.2.2 Formulation of Discrete-Continuous Scheduling Problem with Continuous Resource Discretisation (DCSPwCRD) 131
9.3 Heuristic Algorithms for Solving the DCSP 132
9.4 Metaheuristics for Solving the DCSP 135
9.4.1 TS, SA, and GA as Local Search Metaheuristics for Discrete-Continuous Scheduling Problems 135
9.5 Minimization of the Resource Usage in the DCSP 140
9.6 The Special Case of the DCSP 145
9.7 Research on the Island Model of Computing 146
9.8 Research on Preventing Premature Convergence in Evolutionary and Genetic Algorithms 154
References 156

10 Proposed Metaheuristics for Solving Problem Θ_Z (DCSPwCRD) 161
10.1 IBEA—Island-Based Evolutionary Algorithm 161
10.1.1 Computational Experiment 164
10.2 PLA—Population Learning Algorithm 165
10.2.1 Tabu Search 167
10.2.2 Computational Experiment 168
10.3 PLA2—Cross-Entropy-Based Population Learning Algorithm 170
10.3.1 Cross-Entropy Algorithm 171
10.3.2 Computational Experiment 173

10.4 PLA3—Population Learning with Differential Evolution Algorithm 178
10.4.1 Computational Experiment 180
10.5 IBDEA—Island-Based Differential Evolution Algorithm 183
10.5.1 Computational Experiment 187
References 190

11 Performance Evaluation of the Proposed Algorithms 193
11.1 Friedman Test 194
11.2 Structure Versus Efficiency of the Cross-Entropy-Based Population Learning Algorithm (PLA2) 195
11.2.1 Computational Experiment 200
11.2.2 Conclusions from the Experiment 214
11.3 Properties of the Island-Based and Single Population Differential Evolution Algorithms 215
11.3.1 Computational Experiment 215
11.3.2 Conclusions from the Experiment 219
11.4 Improving Performance of the Differential Evolution Algorithm Using Cyclic Decloning and Changeable Population Size 220
11.4.1 Computational Experiment 221
11.4.2 Decloning Procedure 222
11.4.3 Performance Evaluation Measure 223
11.4.4 Experiments on Decloning 223
11.4.5 Experiments on Population Size 229
11.4.6 Experiments on the Number of Fitness Function Evaluations 229
11.4.7 Performance Improvement Policy 231
11.4.8 Conclusions from the Experiment 232
References 233

12 Conclusions 235

Erratum to: Population-Based Approaches to the Resource-Constrained and Discrete-Continuous Scheduling E1
Ewa Ratajczak-Ropel and Aleksander Skakovski

Acronyms

ABC	Artificial Bee Colony
ABM	Agent-Based Model(s)
ACO	Ant Colony Optimization
AI	Artificial Intelligence
AL	Activity List
AON	Activity On Node
A-Team	Asynchronous Team of Agents
CA	Crossover Algorithm
CAT	Collaborative Agent Team
CE	Cross-entropy algorithm
CFSQP	Specialized solver (A C Code for Solving (Large Scale) Constrained Nonlinear (Minimax) Optimization Problems, Generating Iterates Satisfying All Inequality Constraints)
CGA	Canonical genetic algorithm
CP	Clustering Problem
CPM	Critical Path Method
CPLB	Critical Path Lower Bound
CRSharing	Continuous Resource Sharing Problem
CSM	Cancellation Sequence Method
CT	Computation Time
DAI	Distributed Artificial Intelligence
DCRCPSP	Discrete-Continuous Resource-Constrained Project Scheduling Problem
DCSP	Discrete-Continuous Scheduling Problem
DCSPwCRD	Discrete-Continuous Scheduling Problem with Continuous Resource Discretization
DE	Differential Evolution
DEA	Differential Evolution Algorithm for Solving DCSP
DES	Discrete Event Simulation
DGA	Distributed Genetic Algorithm

DP	Decloning Procedure
DVS	Dynamic Voltage Scaling Scheduling
EA	Evolutionary Algorithm
EDD	Earlier Due Dates
EFT	Earliest Finish Time
EPTSP	Euclidean Planar Traveling Salesman Problem
EST	Earliest Start Time
GA	Genetic Algorithm
GAVRdyskr	GA for Solving DCSPwCRD
HUDD	Heuristic for continuous resource allocation, based on a uniform distribution of the processing demand of tasks
HUDD-PS	HUDD designed for solving DCRCPSP
IBDEA	Island-Based Differential Evolution Algorithm
IBEA	Island-Based Evolutionary Algorithm
JABAT	JADE based A-Team environment
JADE	Java Agent DEvelopement framework
LFT	Latest Finish Time
LLS	Local Left Shift
LPT	Longest Processing Time
LS	Local Search
LSA	Local Search Algorithm
LSG	List-Scheduling Algorithm for Discrete-Continuous Scheduling
LST	Latest Start Time
MCS	Minimal Critical Set(s)
MAS	Multi-Agent System(s)
Max RE	Maximal Relative Error
Mean CT	Mean Computation Time
Mean RE	Mean Relative Error
Mean TCT	Mean Total Computation Time
MPP	Massively Parallel Processor Systems
MRCPSP	Multi-mode Resource-Constrained Project Scheduling Problem
MRCPSP/max	MRCPSP with minimal and maximal time lags
MSIIA	Multi-Start Iterative Improvement Algorithm
NN	Neural-Network
OPL	Optimization Programming Language
opt/bk	optimal or best known (solution)
PBEA	Population-Based Evolutionary Algorithm
PBFS	Percentage of the Best Found Solutions
PERT	Project Evaluation and Review Technique
PGA	Parallel Genetic Algorithm
PLA	Population Learning Algorithm
PLA2	Cross-Entropy-Based Population Learning Algorithm
PLA3	Population Learning with Differential Evolution Algorithm
POS	Potentially Optimal Set
PR	Path-Relinking

PRA	Path-Relinking Algorithm
PS	Population Size
PSO	Particle Swarm Optimization
PSP	Project Scheduling Problem
PSPLIB	Project Scheduling Problems Library
PTA	Precedence Tree Algorithm
RACP	Resource Availability Cost Problem
RCPSP	Resource-Constrained Project Scheduling Problem
RCPSP/max	RCPSP with minimal and maximal time lags
RE	Relative Error
REM	Reverse Elimination Method
RL	Reinforcement Learning
ROG	Random Offspring Generation
RST	Random Sampling Technique
SA	Simulated Annealing
SAA	Version of SA with Optimal Allocation of the Continuous Resource
SAM+	Version of SA for Solving DCSPwCRD
SDT	Social Disasters Technique
SFLA	Shuffled Frog-Leaping Algorithm
SFT	Solution Feasibility Test
SGS	Schedule Generation Scheme
SH	Simple Heuristic
SLA	Social Learning Algorithm
SPP	Scalable Parallel Processor Systems
SPT	Shortest Processing Time
SQP	Sequential Quadratic Programming Method
SS	Scatter Search
TCTO	Time/Cost Trade Off
TCT	Total Computation Time
TL	Tabu List
TNM	Tabu Navigation Method
TS	Tabu Search
TSA	Tabu Search Algorithm
TS-HUDD	HUDD Combined with TS
TS-OPT	TS with Optimal allocation of the continuous resource
VRP	Vehicle Routing Problem
VLSI	Very-Large-Scale Integration

List of Figures

Fig. 2.1 General structure of JABAT . . . 17
Fig. 2.2 Use Case Diagram for JABAT . . . 18
Fig. 4.1 General structure of JABAT solving the RCPSP using SC2 Strategy . . . 42
Fig. 4.2 Pseudocode of the minCT function . . . 44
Fig. 4.3 Pseudocode of the smMakeMove function . . . 45
Fig. 4.4 Pseudocode of the smReverseMove function . . . 45
Fig. 4.5 Pseudocode of the smMakeExchange function . . . 45
Fig. 4.6 Pseudocode of the smReverseExchange function . . . 45
Fig. 4.7 Pseudocode of the smLSAm algorithm . . . 46
Fig. 4.8 Pseudocode of the smTSAe algorithm . . . 47
Fig. 4.9 Pseudocode of the smCA algorithm . . . 47
Fig. 4.10 Pseudocode of the smPRA algorithm . . . 48
Fig. 4.11 General schema of the DCPL Strategy . . . 54
Fig. 4.12 General schema of the DCI Strategy . . . 58
Fig. 4.13 Graphical representation of the results from Table 4.9 . . . 59
Fig. 4.14 The mean values of Friedman test ranks for the RCPSP . . . 60
Fig. 5.1 Pseudocode of the mmMakeMove function . . . 80
Fig. 5.2 Pseudocode of the mmReverseMove function . . . 80
Fig. 5.3 Pseudocode of the mmMakeExchange function . . . 80
Fig. 5.4 Pseudocode of the mmReverseExchange function . . . 80
Fig. 5.5 Pseudocode of the mmLSAm algorithm . . . 81
Fig. 5.6 Pseudocode of the mmTSAe algorithm . . . 82
Fig. 5.7 Pseudocode of the mmCA algorithm . . . 83
Fig. 5.8 Pseudo-codes of the mmPRA algorithm . . . 83
Fig. 5.9 Graphical representation of the results from Table 5.9 . . . 90
Fig. 5.10 The mean values of Friedman test ranks for the MRCPSP . . . 92
Fig. 8.1 The division of a feasible schedule into intervals M_k defined by the completion times of consecutive tasks . . . 116
Fig. 8.2 The division of processing demands of tasks $\tilde{x}_i$ into parts $\tilde{x}_{ik}$, corresponding to time intervals M_k . . . 116

Fig. 9.1 The 6 migration topologies considered in [62] 150
Fig. 11.1 The means of ranks obtained for the metaheuristics under the test 195
Fig. 11.2 A simplified scheme of SO algorithm based on ring topology 199
Fig. 11.3 A simplified scheme of SX algorithm based on random topology 199
Fig. 11.4 A simplified scheme of AO algorithm based on ring topology 200
Fig. 11.5 A simplified scheme of AX algorithm based on random topology 200
Fig. 11.6 The results obtained for the test sets 1–137 217
Fig. 11.7 The results obtained for the test sets 92–137 218
Fig. 11.8 The effect of decloning on $sumC_{max}$, $x_P = 20$, $\#ev = 37800$, $T^d \in [20, 37800]$ 218
Fig. 11.9 $sumC_{max}$ yielded by the DEA without and with decloning, $x_P = 20$, $\#ev = 37800$, $T^d = 20$ 224
Fig. 11.10 The effect of decloning on $sumC_{max}$, $x_P = 20$, $\#ev = 720000$, $T^d \in [20, 238200]$ 225
Fig. 11.11 The effect of decloning on $sumC_{max}$, $x_P = 50$, $\#ev = 720000$, $T^d \in [50, 249200]$ 225
Fig. 11.12 The effect of decloning on $sumC_{max}$, $x_P = 100$, $\#ev = 720000$, $T^d \in [100, 249200]$ 226
Fig. 11.13 The effect of decloning on $sumC_{max}$, $x_P = 200$, $\#ev = 720000$, $T^d \in [200, 364200]$ 226
Fig. 11.14 The effect of decloning on $sumC_{max}$, $x_P = 1000$, $\#ev = 720000$, $T^d \in [1000, 250000]$ 227
Fig. 11.15 The improvement of the results due to decloning for different population sizes x_P compared to the case without decloning, given in percent 227
Fig. 11.16 The difference between AVG $sumC_{max}$ values obtained by the DEA with and without decloning for population sizes $x_P = 20$, $x_P = 50$, $x_P = 1000$ 228
Fig. 11.17 The difference between AVG $sumC_{max}$ values obtained by the DEA with and without decloning for population sizes $x_P = 100$, $x_P = 200$ 229
Fig. 11.18 AVG $sumC_{max}$ of the DEA without decloning, considered for different x_P and $\#ev$ 230
Fig. 11.19 The effect of decloning on AVG $sumC_{max}$, considered for different x_P and $\#ev$ 231

Part I
Agent-Based Approach to the Single and Multi-mode Resource-Constrained Project Scheduling

Ewa Ratajczak-Ropel

Chapter 1
Introduction

The Resource-Constrained Project Scheduling Problem (RCPSP), as well as its numerous extensions and special cases, have attracted a lot of attention and many exact, heuristic and metaheuristic solution methods have been proposed in the literature in recent years [1–4]. Due to the problem complexity the current approaches to solving instances of the discussed class produce either approximate solutions or can be applied to solving instances of only limited size. Hence, searching for more effective algorithms and solutions to the problem is still a lively field of research. One of the promising directions of such research is to take advantage of the parallel and distributed computations, which are common features of the contemporary multiagent systems [5].

Agent-based approaches or agent-based computing are most frequently referred to as Multi-Agent Systems (MAS) or Agent-Based Models (ABM) [6], which are a subfield of the Distributed Artificial Intelligence (DAI). DAI is, in turn, a subfield of the Artificial Intelligence (AI). MAS have been studied as an autonomous field of research since about 1980s and gained widespread recognition in the mid 1990s. Since then, agent-based approaches are established as an important and intensively expanding area of research and development. The state of the art review related to the agent approaches can be found, for example in [5–13]. In the literature different aspects of agent-based approaches are widely considered, for example: machine learning [14], cooperative multi-agent learning [15], agents communities [16], privacy [17], and agent platforms [18].

Multi-agent systems [5] deal with the behavior of the computing entities available to solve together a given problem. In MAS each computing entity is referred to as an agent. Multi-agent system can be defined as a network of individual agents which have at least two important capabilities. They are capable of acting autonomously and can interact with each other in order to solve a problem.

E. Ratajczak-Ropel and A. Skakovski, *Population-Based Approaches to the Resource-Constrained and Discrete-Continuous Scheduling*,
Studies in Systems, Decision and Control 108, DOI 10.1007/978-3-319-62893-6_1

Multi-agent systems can be used to solve problems that are too difficult or impossible to deal with by an individual agent or a monolithic system. They may include some functional or procedural approaches, algorithmic search or reinforcement learning algorithms. Multi-agent systems are widely used to solve different optimization problems, including project scheduling. The survey of the agent-based approaches to optimization can be found, for example, in [12, 19, 20]. Examples of using MAS for solving optimization problems include: scheduling problems [21, 22], vehicle routing problems [23, 24], supply chain management [25], traffic management [26, 27] and many others.

Modern Multi-Agent System architectures are an important and intensively expanding area of research and development. There exists a number of multiple-agent approaches proposed to solve different types of optimization problems. One of them is the concept of the A-Team, originally introduced in [28]. The idea of the A-Team was used to develop a software environment for solving a variety of computationally hard optimization problems called JADE based A-Team (JABAT) [29, 30]. JABAT system supports construction of the dedicated A-Team architectures. Agents used in JABAT assure decentralization of computations across multiple hardware platforms. Parallel processing results in more effective use of the available resources and ultimately, in reduction of the computation time.

In this part of the book the classical Resource-Constrained Project Scheduling Problem (RCPSP) and its generalization Multi-mode Resource-Constrained Project Scheduling Problem (MRCPSP) are considered. The state-of-the-art review of the agent-based solution approaches to these problems is presented, including author's results. Agent-based algorithms for solving the above mentioned problems are described and experimentally validated using problem instances from PSPLIB library [31]. Solutions are compared with these known from the literature.

The first part of the book is constructed as follows. Chapter 2 contains some general ideas and the literature review of agent-based optimization. Chapter 3 reviews the project scheduling models. Chapter 4 focuses on the resource-constrained project scheduling, including state-of-the-art review and the proposed approaches to solving the RCPSP. In Chap. 5 the multi-mode resource-constrained project scheduling is shortly reviewed and several approaches to solving it using the agent-based paradigm are proposed and evaluated experimentally. Finally, Chap. 6 contains conclusions and short discussion of open problems. Research results obtained by the author with co-authors and published in a number of papers concerning the A-Team approaches to solving the family of the resource-constrained project scheduling problems, in particular RCPSP and MRCPSP are presented in Sects. 4.4 and 5.4.

References

1. Kölisch, R., Hartmann, S.: Experimental investigation of heuristics for resource-constrained project scheduling: an update. Eur. J. Oper. Res. **174**(1), 23–37 (2006)
2. Agarwal, A., Colak, S., Erenguc, S.: A neurogenetic approach for the resource-constrained project scheduling problem. Comput. Oper. Res. **38**, 44–50 (2011)
3. Paraskevopoulos, D.C., Tarantilis, C.D., Ioannou, G.: Solving project scheduling problems with resource constraints via an event list-based evolutionary algorithm. Expert Syst. Appl. **39**, 3983–3994 (2012)
4. Fang, C., Wang, L.: An effective shuffled frog-leaping algorithm for resource-constrained project scheduling problem. Comput. Oper. Res. **39**(5), 890–901 (2012)
5. Wooldridge, M.: An Introduction to Multiagent Systems, 2nd edn. Wiley (2009)
6. Niazi, M., Hussain, A.: Agent-based computing from multi-agent systems to agent-based models: a visual survey. Scientometrics **89**(2), 479–499 (2011)
7. Wooldridge, M., Jennings, N.R.: Intelligent agents: theory and practice. Knowl. Eng. Rev. **10**(2), 115–152 (1995)
8. Wooldridge, M.: An Introduction to Multiagent Systems. Wiley, New York (2002)
9. Jennings, N.R., Wooldridge, M.: Applying agent technology. Appl. Artif. Intell. **9**, 357–369 (1995)
10. Moulin, B., Chaib-Draa, B.: An overview of distributed artificial intelligence. In: O'Hare, G.M.P., Jennings, N.R. (eds.) Foundations of distributed artificial intelligence, pp. 3–55. Wiley, New York (1996)
11. Balaji, P.G., Srinivasan, D.: An introduction to multi-agent systems. In: Innovations in Multi-agent Systems and Applications—1, Studies in Computational Intelligence, vol. 310, pp. 1–27. (2010)
12. Barbati, M., Bruno, G., Genovese, A.: Applications of agent-based models for optimization problems: a literature review. Expert Syst. Appl. **39**, 6020–6028 (2012)
13. Naciri, N., Tkiouat, M.: Multi-agent systems: theory and applications survey. Int. J. Intell. Syst. Technol. Appl. **14**(2), 145–167 (2015)
14. Stone, P., Veloso, M.: Multi-agent systems: a survey from a machine learning perspective. Auton. Robot. **8**(3), 345–383 (2000)
15. Panait, L., Luke, S.: Cooperative multi-agent learning: the state of the art. Auton. Agent. Multi-Agent Syst. **11**(3), 387–434 (2005)
16. Michel, F., Ferber, J., Drogoul, A.: Multi-agent systems and simulation: a survey from the agent community's perspective. In: Multi-Agent Systems: Simulation and Applications, pp. 3–52. CRC Press, Boca Raton, FL (2009)
17. Such, J.M., Espinosa, A., García-Fornes, A.: A survey of privacy in multi-agent systems. Knowl. Eng. Rev. **29**(03), 314–344 (2014)
18. Kravari, K., Bassiliades, N.: A survey of agent platforms. J. Artif. Soc. Soc. Simul. **18**(1), 11 (2015)
19. Persson, J.A., Davidsson, P., Johansson, S.J., Wernstedt, F.: Combining agent-based approaches and classical optimization techniques. In: Proceedings of the Third European Workshop on Multi-Agent Systems (EUMAS 2005), pp. 260–269 (2005)
20. Ren, H., Wang, Y.: A survey of multi-agent methods for solving resource constrained project scheduling problems. In: Proceedings of International Conference on Management and Service Science 2011, pp. 1–4 (2011)
21. Knotts, G., Dror, M.: Agent-based project scheduling: computational study of large problems. IIE Trans. **35**, 143–159 (2003)
22. Aydin, M.: Metaheuristic agent teams for job shop scheduling problems. In: Holonic and Multi-Agent Systems for Manufacturing. Lecture Notes in Computer Science, vol. 4659, pp. 185–194. (2007)
23. Barbucha, D., Jędrzejowicz, P.: An agent-based approach to vehicle routing problem. Int. J. Appl. Math. Comput. Sci. **4**(2), 538–543 (2007)

24. Xie, X.F., Liu, J.: Multiagent optimization system for solving the traveling salesman problem (TSP). IEEE Trans. Syst. Man Cybern. Part B Cybern. **39**(2), 489–502 (2009)
25. Liang, W.Y., Huang, C.C.: Agent-based demand forecast in multi-echelon supply chain. Decis. Support Syst. **42**(1), 390–407 (2006)
26. Blum, J., Eskandarian, A.: Enhancing intelligent agent collaboration for flow optimization of railroad traffic. Transp. Res. **36**(10), 919–930 (2002)
27. Chen, B., Cheng, H.H., Palen, J.: Integrating mobile agent technology with multi-agent systems for distributed traffic detection and management systems. Trans. Res. Part C Emerg. Technol. **17**(1), 1 (2009)
28. Talukdar, S., Baerentzen, L., Gove, A., De Souza, P.: Asynchronous teams: co-operation schemes for autonomous, computer-based agents. Technical Report EDRC 18-59-96, Carnegie Mellon University, Pittsburgh (1996)
29. Jędrzejowicz, P., Wierzbowska, I.: JADE-Based A-Team environment. In: Computational Science—ICCS. Lecture Notes in Computer Science, vol. 3993, pp. 719–726 (2006)
30. Barbucha, D., Czarnowski, I., Jędrzejowicz, P., Ratajczak-Ropel, E., Wierzbowska, I.: e-JABAT—an implementation of the web-based A-Team. In: Nguyen, N.T., Jain, L.C. (eds.) Intelligence Agents in the Evolution of Web and Applications. Studies in Computational Intelligence, vol. 167, pp. 57–86. (2009)
31. PSPLIB - Project Scheduling Problem LIBrary. http://www.om-db.wi.tum.de/psplib

Chapter 2
Agent-Based Optimization

Agents and agent-based approaches are an active research topics in artificial intelligence and expert systems. Due to their properties, they are recently being used as a promising tool for solving problems whose domains are distributed, complex and hetergenous.

In this chapter the short overview of the agent-based applications for solving optimization problems is presented. In Sect. 2.1 basics of the agent-based approaches are introduced. Agent-based approaches to optimisation are described in Sect. 2.2. In Sect. 2.3 agent-based approaches to project scheduling, including the A-Team paradigm are presented.

2.1 Basics of the Agent-Based Approaches

The term “agent”, or software agent, has found its way into a number of technologies and has been widely used, for example, in artificial intelligence, databases, operating systems and computer networks literature. Although there is no single definition of an agent in the literature [1–3], all definitions agree that an agent is essentially a special software component that has autonomy, that provides an interoperable interface to an arbitrary system and/or behaves like a human agent, working for some clients in pursuit of its own agenda [4].

Agent-based approaches form a large and widely spread research domain ranging from the computer science techniques for constructing and implementing agent-based systems on one side, to modeling concepts taking their roots in social and natural sciences, on the other. In the literature diverse concepts and terms are considered such as agent-based computing, agent-based modeling and simulation, agent-based software engineering or multi-agent systems [5, 6]. Term “agent” may cover a range of entities from a rather simple software agents or services/deamons, which might

E. Ratajczak-Ropel and A. Skakovski, *Population-Based Approaches to the Resource-Constrained and Discrete-Continuous Scheduling*, Studies in Systems, Decision and Control 108, DOI 10.1007/978-3-319-62893-6_2

not behave very intelligently to intelligent agents. The latter use artificial intelligence concepts and methods to control their behavior [7, 8]. An agent could be also an interacting social component of a large system used to explore new global behavior in a simulation experiments [6].

In the most general way in which the term "agent" is used, it is understood as computational system. According to the definition of Wooldridge and Jennings [2] an agent is a computational system interacting with an environment that can be endowed with the following features:

- autonomy - each agent acts without the direct control of human beings or other devices;
- social ability - interactions occur among entities through a communication language in order to satisfy the objectives;
- reactivity - agents answer in a precise way to signals coming from the environment;
- pro-activeness - agents do not simply act in response to their environment, they take the initiative in order to satisfy their goal.

Agent-Based Models (ABM) and Multi-Agent Systems (MAS) consist of a set of elements (agents) characterized by some attributes, which interact with each other through the definition of appropriate rules in a given environment. A slight difference between ABM and MAS is frequently indicated.

A Multi-Agent System (MAS) is a computerized system composed of multiple intelligent agents interacting within an environment, while an Agent-Based Model (ABM) is one of the class of computational models. ABMs are used for simulating actions and interactions of autonomous agents (both individual or collective entities such as organizations or groups) with a view to assessing their effects on the system as a whole. ABM combines elements of game theory, complex systems, emergence, computational sociology, multi-agent systems, and evolutionary programming. The goal of an ABM is to search for explanatory insight into the collective behavior of agents obeying simple rules, typically in natural systems, rather than solving specific practical or engineering problems. The terminology of the ABM tends to be used more often in the sciences, and MAS in engineering and technology.

In [9] Madejski indicates that there are two different approaches to agent design: the physical decomposition approach and the functional decomposition one. In the first case, agents represent physical entities, like workers, machines, tools, fixtures, or products, etc. On the other hand, in the functional decomposition approach, there is no relationship between agents and physical entities, but agents are assigned to some functions like product distribution, transport management, order acquisition, scheduling, material handling, etc.

Multi-agent systems may be described and classified based on several different attributes, such as: architecture, learning and decision making abilities, communication, coordination etc. Recent review of this topic can be found in [10], in which the following features are considered:

- Internal architecture
  - Homogeneous structure
  - Heterogeneous structure
- Overall Agent Organization
  - Hierarchical Organization
  - Holonic Agent Organization
  - Coalitions
  - Teams
- Communication
  - Local Communication
  - Blackboards
  - Agent Communication Language
- Decision making
  - Nash equilibrium
  - Iterated elimination method
- Coordination
  - Coordination through protocol
  - Coordination via graphs
  - Coordination through belief models
- Learning
  - Active Learning
  - Reactive Learning
  - Learning based on consequences

Multi-agent systems are widely applied in many domains because of the beneficial advantages offered. Some of the benefits of using MAS technology in large systems are indicated in [10, 11]:

- speedup and efficiency - due to the asynchronous and parallel computations;
- robustness and reliability - the whole system can undergo a 'graceful degradation' when one or more agents fail;
- scalability and flexibility - it is easy to add new agents to the system;
- cost - an agent is a low-cost unit compared to the whole system;
- development and reusability - it is easier to develop and maintain a modular system than a monolithic one.

Agent approaches are applied to solve different optimization problems. Description of the most recent trends in the field of agent-based optimization and decision support techniques can be found in [12]. Therefore, multi-agent systems are used in a variety of applications, ranging from comparatively small systems for personal assistance to open, complex, mission critical systems for industrial applications [13].

The first MAS techniques were developed alongside with industrial applications, e.g. process control [14], manufacturing, system diagnostics [15], transportation logistics [16], or network management [17]. Other application fields of multi-agent systems include: information management, traffic and transportation, telecommunication systems, computer games, graphics and health care.

2.2 Agents-Based Approaches to Optimization

In recent years numerous papers proposing ABM or ABM-based approaches to optimization have been published. Usually, such approaches combine agent-based paradigms with other optimization techniques, distributed or complex systems, heuristic methods, management sciences etc. Johnson et al. [18] distinguish three most common forms of integration of the optimization and agent-based models:

- optimization used as a calibration and validation tool for ABM,
- ABM used to solve optimization problems,
- optimization used in economic ABM to represent constrained maximisation.

In this chapter the second form is considered where ABM is used as a method, technique or framework for solving optimization problems. There is a number of multiple-agent approaches proposed in the literature to solve different types of optimization problems. However, only a few papers present surveys extended literature reviews, or comparisons of these methods.

In the field of intelligent manufacturing the idea of two different categories of agents with different roles with regard to optimization, emerged in [9]. Agents of the first category represent or directly supply the required resources within optimization effort. Agents of the second category perform certain functions. The first category includes physical agents, representing physical entities such as workers, vehicles, products, machines, resources or users. Examples include [19–22]. The second category, functional agents, represent pieces of software used to carry out subtasks such as search or working strategies, local optimization or project management. Examples include solutions presented in [23–26] as well as Sects. 4.4 and 5.4.

Barbati et al. [27] review several approaches to using agents for solving optimization problems. They also identify two agent-based architectures often used to solve scheduling optimization problems: autonomous and mediator. In the autonomous architecture many agents self-organize themselves to solve a problem based on negotiation protocols. MAS is composed of autonomous cooperating local agents that are capable of negotiating with each other in order to achieve their aims (see for example [19, 22, 26, 28–30]). In the mediator architecture based on cooperative interaction protocols, a mediator is responsible for the coordination of agents objectives, in such a way that the quality of the global solution is continuously improved. Mediator agents coordinate the behavior of the local agents to perform global optimization. Examples can be found in [23, 25, 31–33]. Classification of architectures proposed in [27] is general and covers majority of approaches to solving optimization problems through employing various types of agents.

Persson et al. [34] and Davidsson et al. [32] compare strengths and weaknesses of agent-based approaches versus classical optimization techniques. They focus on evaluating how well both approaches are able to handle some important properties of the problem domain. Authors propose a set of properties enabling comparison of agent-based approaches and classical optimization techniques, including:

- size (number of resources to be allocated),
- cost of communication,
- communication and computational stability,
- modularity,
- time scale (time between re-allocation of resources),
- changeability (how often the structure of the domain changes),
- quality of solution (how important it is to find a good allocation),
- quality assurance,
- integrity (importance of not distributing sensitive information).

Based on this analysis it is indicated that the properties of the agent-based approaches and optimization techniques complement each other and there exists a number of ways for combining them. They indicate and describe two such approaches:

- Assisting agents with a plan obtained through using optimization techniques. The approach can be understood as using some optimization technique for coarse planning and agents for operational replanning, i.e., for performing local adjustments of the initial plan in real-time to handle the actual conditions when the plan is executed.
- Embedding optimization capabilities within agents. In the deployed distributed systems this requires the use of wrapper technology, or a similar solutions, in order to make an agent a fully integrated first-class citizen of the multi-agent system.

The above briefly reviewed papers illustrate current state of the art related to the use and to the application of agent-based models as optimization tools. One of such tools is the concept of an A-Team, originally introduced by Talukdar et al. [35, 36].

The idea of the A-Team has been used to develop the software environment for solving a variety of computationally hard optimization problems called JABAT [37, 38]. JADE based A-Team (JABAT) system supports the construction of the dedicated A-Team architectures. Agents used in JABAT assure decentralization of computation across multiple hardware platforms. Parallel processing results in more effective use of the available resources and ultimately, a reduction of the computation time. In the following section the concept of the A-Team and JABAT are introduced. Several applications of this system to implement A-Teams solving variety of scheduling problems are described in Sects. 4.4 and 5.4.

2.2.1 A-Team Concept

The idea of asynchronous agents working as a team in order to solve optimization problem has been proposed and developed by Talukdar et al. in [35, 36, 39, 40], and then developed and used by another authors, for example in [38, 41–44].

According to [39] an asynchronous team is a collection of software agents that cooperate to solve a problem by dynamically evolving a population of solutions. Formal definition of the A-Team has been proposed in [35, 36]: an A-Team is seen as a set of autonomous agents and a set of memories, interconnected to form a strongly cyclic computational network, that is, a network in which every agent is in the closed loop. An A-Team can be visualized as a directed hypergraph, e.g. a data flow where each node represents a complex of overlapping memories, and each arc represents an autonomous agent. The results or trial-solutions are stored in the memories, similarly to blackboard systems, and form the population of solutions. The role of such populations is similar to the populations in genetic algorithms: new individuals are continually added by the construction agents, while other are being erased by the destruction agents.

In the later papers of Talukdar et al. [40] A-Team is described as a multi-population, multi-agent system for solving optimization problems. In the A-Team, search of solution skills are packaged as agents. During the computation process problem is decomposed into sub-problems and a population of solutions is maintained for each sub-problem. These populations are iteratively evolved by a set of agents. Additionally, solutions are circulating among populations. Some solutions in each population will eventually improve. The computation process is terminated when further iterations do not bring any improvement to the so far obtained best solution.

The function of the A-Team is to combine operators or algorithms, so together they can tackle bigger and more difficult problems than they could if working alone [35]. Hence, the collaboration between agents are particularly important. In [45] Talukdar has proposed a grammar for constructing asynchronous teams that might be useful in solving an instance of the off-line problems. In another words, the grammar constructively defines the space that must be searched if an asynchronous team that is good at solving the given problem-instance is to be found. The most important part of this grammar includes principles and rules for designing problem-solving organizations in which collaboration among such agents is automatic and scale-effective. The primitives of the grammar are:

- sharable memories, each dedicated to a member of the family-of-problems, and designed to contain a population of trial-solutions to its problem;
- operators for modifying trial-solutions;
- selectors for picking trial-solutions;
- schedulers for determining when selectors and operators are to work.

The rules of the grammar are:

- Form autonomous agents by packaging an operator with a selector and a scheduler.
- Use quality-based-selection and completely parallel execution (all the agents running all the time, or as close to all the time, as the available computer resources will allow) as the default selection and scheduling strategies.
- Connect the agents and memories to form a strongly cyclic data flow.
- Compensate for construction deficiencies with skilled destruction.
- Mix agents as needed without regard to their complexity or phylla, that is big and small software agents may be combined with humans, provided only that the humans subscribe to the communication and selection conditions prescribed for the software agents.

According to Correa et al. [41] the execution of the A-Team can be described by the set of events. Each event is composed from the following elements: the time at which the event occurs; the input data, if any; the state of the shared memory prior to the occurrence of the event; the state of the shared memory after the occurrence of the event; and the output data, if any. There is a restricted number of types of events occurring during the A-Team execution:

- initialization of the global variables and the shared memory,
- reading operation,
- execution of the heuristic/search,
- writing operation.

Implementations and applications of the A-Team for solving optimization problems are described and reviewed in several papers. A-Team based architectures, systems, environments, frameworks and metaheuristics are developed for a wide variety of optimization problems. The reviews of these approaches can be found in [43, 44, 46].

Rachlin et al. in [46] propose and use A-Team architecture to develop real-world optimization and decision support applications. The proposed implementation provides the basic components needed to create A-Teams, the configuration language for assembling and customizing components and the user interface for interfacing with the resulting A-Team. In the same paper authors present the A-Team overview and indicate some important features of such architectures. They also outline key advantages of the approach: modularity, suitability for distributed environments and robustness.

Jędrzejowicz in [43] divides the A-Team implementations into first and next generation. The major differences can be found with respect to accessibility, scalability and portability. In the same paper two major classes of the A-Team implementations are defined. The first includes specialized A-Teams designed to solve instances of particular problems. Their architecture is problem-specific and not flexible. The second class covers middleware platforms allowing for an easy implementation of the A-Teams ready to solve instances of the arbitrary problems.

Carle et al. in [44] propose the metaheuristic based on the A-Team paradigm designed to tackle complex multi-dimensional optimization problems called

Collaborative Agent Team (CAT). In the paper a discussion about designing three components of CAT for a particular optimization problem can be found: the problem representation along different dimensional views, the design of the agents and the information sharing between them.

In [27, 38, 47] the A-Team based environment for solving different optimization problems called JADE-based A-Team (JABAT) has been proposed and experimentally validated. Because JABAT has been used to implement approaches proposed in this book, it is described separately in the following subsection.

A-Team based approaches for solving optimization problems include: traveling salesman problems [39], control of electric networks [48, 49], collision avoidance in robotics [50], planning and scheduling in manufacturing [51], flow optimization of railroad traffic [52], job-shop scheduling [53–55], steel and paper mill scheduling [56, 57], train scheduling [58], automatic insertion of electronic components [59], non-fixed point-to-point connection problem [41], clustering problem [60], euclidean planar traveling salesman problem [61], multi-period supply chain network design problem [62], vehicle routing problem [63], and vehicle routing problem with time windows [64, 65].

2.2.2 A-Team Implementation - JABAT

One of implementations of the A-Team concept is JADE-based A-Team Environment called JABAT. It has been implemented and developed by the team of researchers from Department of Information Systems in Gdynia Maritime University with the participation of the author [38, 47, 66]. The platform offers a generic A-Team architecture allowing users to execute different population-based methods with some default and/or user-defined optimization procedures implemented as agents within the asynchronous team of agents.

JABAT is a middleware supporting the construction of the dedicated A-Team architectures used for solving different computationally hard optimization problems. JABAT engine is JADE, which is based on Java technologies. To construct JABAT also Java technologies have been used, including Java 2 Platform Standard Edition (J2SE) with Java Runtime Environment (JRE).

JADE (Java Agent DEvelopement framework) [67, 68] is an enabling technology for the development and run-time execution of peer-to-peer applications which are based on the agents paradigm and which can seamlessly work and interoperate both in wired and wireless environment. JADE is best described as the distributed middleware system, multi-agent system, software framework or FIPA-compliant agent platform. It facilitates the development of agent-based applications through the runtime environment, the core logic of agents, and a graphical tools. It has a flexible infrastructure allowing easy extension with add-on modules. As it is written in Java, it benefits from a huge set of Java language features and third-party libraries [4]. From the functional point of view, JADE provides the basic services necessary to construct the distributed peer-to-peer applications in the fixed and mobile environment. JADE allows each agent to dynamically discover other agents and to communicate within the team according to the peer-to-peer paradigm.

The problem-solving engine on which JABAT is based can be best defined as the population based approach. The environment is expected to be able to produce solutions to difficult optimization problems through applying the following general rules [38]:

- To solve difficult optimization problems use a set of agents, each representing an improvement algorithm.
- To escape getting trapped into a local optimum generate or construct the initial population of solutions called individuals, which, during computations will be improved by agents, thus increasing chances for reaching the global optimum.

Agent-based architecture of JABAT allows implementation of the following features [38, 69]:

- The system can in parallel solve instances of several different optimization problems.
- A user, having a list of all algorithms implemented for the given problem, may choose how many and which of them should be used.
- The optimization processes can be performed on many computers. The user can easily add or delete a computer from the system. In both cases JABAT will adapt to the changes, commanding the agents working within the system to migrate.
- The system is fed in the batch mode - consecutive problems may be stored and solved later, when the system or user assesses that there is enough resources to undertake a new search.

JABAT produces solutions to combinatorial optimization problems using a set of optimization agents. Each agent represents an improvement algorithm. An initial population of solutions (individuals) is generated or constructed using, for example, some heuristics. Individuals from the population are, at the following computation stages, improved by independently acting agents. Main functionality of JABAT includes organizing and conducting the process of search. It involves a sequence of the following steps [47]:

- Generating an initial population of solutions.
- Applying solution improvement algorithms which draw individuals from the common memory and store them back after attempted improvement, using some user defined replacement strategy.
- Continuing reading-improving-replacing cycle until a stopping criterion is met.

The JABAT environment is based on two main agents types and three types of special agents. All agents are implemented as Java classes. The main agents are TaskManager and PlatformManger which manage all other agents and hardware platforms. Both main agents are running continuously on the main platform placed on a server. The special agent types are SolutionManager, SolutionMonitor and OptiAgent. The OptiAgent represents the optimization algorithm which is used to solve some particular optimization problem. The OptiAgent class must be overwritten by the code specifically designed for solving a particular problem type. The SolutionManager manages the population of solutions. Its class may be overwritten to implement a specific user designed cooperation strategy with respect to handling the population of solutions. Such a strategy defines which solutions and when are deleted from the common memory, how they are replaced and how and when new individuals are generated and incorporated into the common memory. The SolutionMonitor is responsible for registering solutions obtained by OptiAgents. The class may be overwritten to make possible recording partial results of computations. One SolutionManager, one SolutionMonitor and a number, fixed or variable, of OptiAgents are run for each problem instance. The general structure of JABAT is shown in Fig. 2.1. The Use Case Diagram for JABAT is shown in Fig. 2.2. The detailed description of the JABAT environment may be found in [38].

Apart from the above described JABAT-specific classes several general classes describing a particular optimization problem need to be defined. They include:

- Data - representing problem data and using a set of text files describing instances of the considered problem,
- Task - representing an instance of the problem saved in Java structures,
- Solution - representing a solution of the problem.

For each of the mentioned classes respective ontology class has been implemented. They include: DataOntology, TaskOntology and SolutionOntology. Implementing the JABAT code specific for the considered problem requires overwriting all of these classes.

JABAT has two extensions:

- e-JABAT - implementation of Web-based A-Team which are fully Internet accessible, portable, scalable and in conformity with the FIPA standards proposed in [37].
- TA-Teams JABAT - implementation of Team of A-Teams in which several A-Teams work in parallel and cooperate to solve optimization problems [70].

JABAT, as well as its extensions has been successfully used for implementing dedicated A-Team architectures solving different NP-hard optimization problems. These problems include: Euclidean Planar Traveling Salesman Problem (EPTSP) [61], Vehicle Routing Problem (VRP) [63], Clustering Problem (CP) [60], Resource Availability Cost Problem (RACP) [71], as well as the single and multi-mode resource-constrained project scheduling problems described in this book: RCPSP and MRCPSP.

To implement dedicated A-Team architecture in the JABAT environment it is necessary to construct three sets of classes/agents:

- base classes - representing instance data, solution and the respective ontologies;
- optimization agents - classes representing optimization algorithms, each algorithm is build in one agent class;
- strategies - classes describing strategies used to manage the process of problem solving.

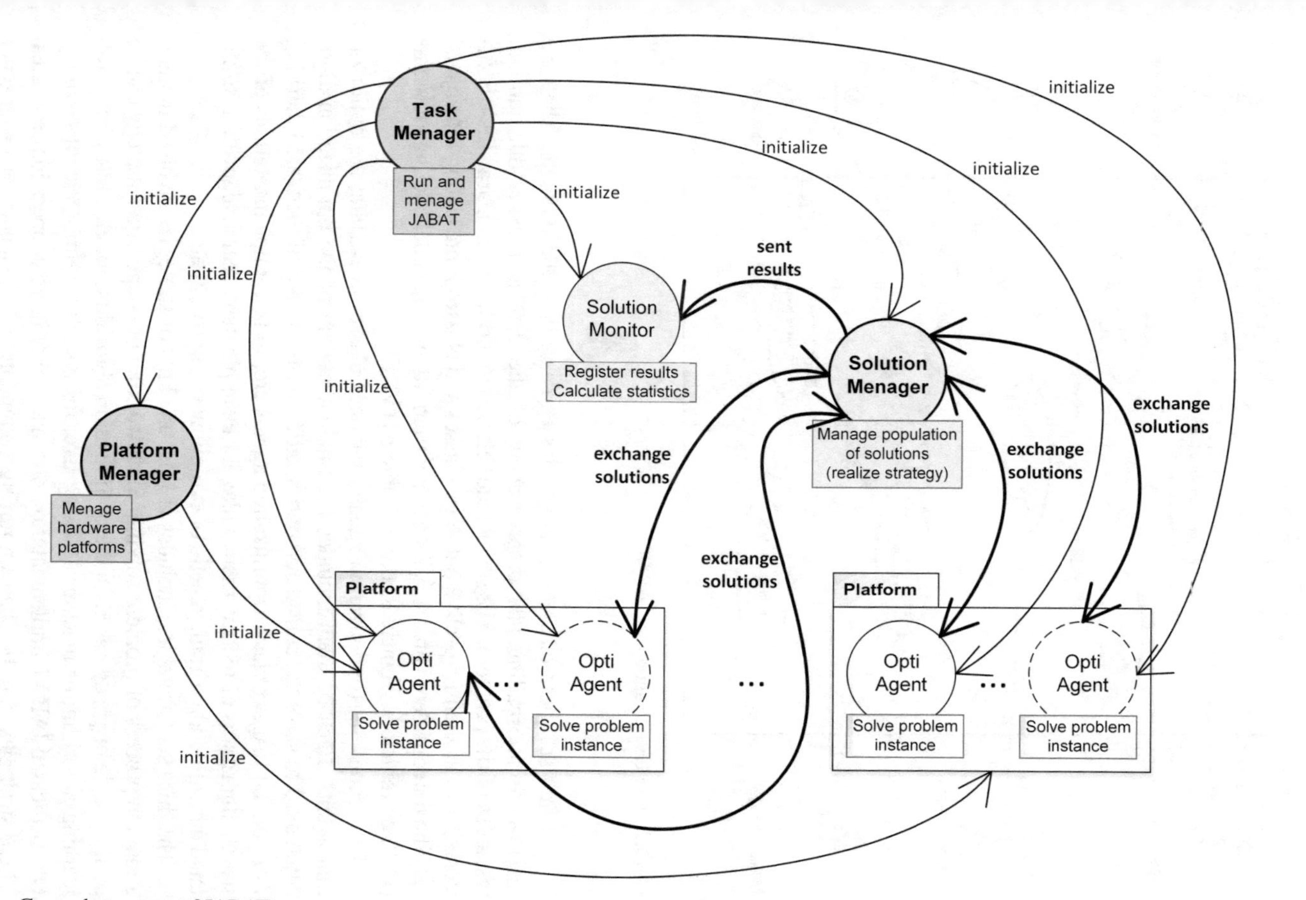

Fig. 2.1 General structure of JABAT

Fig. 2.2 Use Case Diagram for JABAT

The first set (base classes) includes classes describing the problem. They are responsible for reading and preprocessing of the data and generating random instances of the problem. The set includes classes inheriting from Data, Task, Solution as well as their respective ontology classes and classes representing the objects specific for considered problem, for example activity in Project Scheduling Problems (PSP) or vehicle in Vehicle Routing Problems (VRP).

The second set (optimization agents) includes classes describing the optimization agents. Each of them includes the implementation of the optimization algorithm used to solve the considered optimization problem. All of them are inheriting from the OptiAgent class. Optimization agents are, in fact, implementations of the search algorithms or metaheuristics like, for example: tabu search algorithm, simulated annealing algorithm, genetic or evolutionary algorithm etc.

The third set (strategies) includes at least one kind of strategies involved in managing the process of problem solving. Such working or cooperation strategy can be on the whole or partly based on the optimization algorithm, for example population learning algorithm or machine learning like, for instance, reinforcement learning. One dedicated JABAT implementation may use one or more cooperation strategies. These strategies may be changed during computation in sequence or using other arrangement. For example in the JABAT extension called TA-Teams the second kind of strategy is used. It manages the cooperation/interactions between A-Teams. It is called interaction strategy [61, 72, 73].

2.3 Agents-Based Approaches to Project Scheduling

Majority of approaches proposed for optimization, has been successfully adopted to scheduling and project scheduling. Project scheduling can be classified according to the number of projects as single-project scheduling or multi-project scheduling. Otherwise, in regard to the methods of management the static and dynamic scheduling as well as centralized and distributed scheduling can be considered.

Contemporary agent approaches can be classified taking into consideration the agents role in the system. Most frequently, the agents representing the problem features are used, so the following categories are considered:

- Activity as agent approach - each activity of the project is represented as simple agent.
- Resource as agent approach - each resource of the project is represented as a single agent.
- Project as agent approach - each agent represents the project as a whole, or part of the project.
- Process as agent approach - each agent represents a process carried-out within the system. For example an agent may represent the resource allocation process.

The above listed categories do not exclude each other. For example in the systems where activity as agent approach is used activity as resource approach can be used simultaneously. Contemporary MAS used in project scheduling belong usually to activity and resource as agents categories or project as agent category. Additionally, other components of the system can be represented by agents, for example memory, optimization or learning algorithms, etc.

The short survey of some agent-based scheduling approaches can be found in [74], including dynamic, distributed, multi-mode and multi-project problems.

Different multi-agent-based optimization models, systems and frameworks have been proposed for project management problems in the literature. The reviews of such approaches can be found in [28, 74].

Jędrzejowicz and Ratajczak-Ropel in [75–82] have proposed several models based on the asynchronous team of agents (A-Team) solving some project scheduling problems using JABAT environment described in Sect. 2.2. Details of some of these approaches are given in Chaps. 4 and 5.

References

1. Genesereth, M.R., Ketchpel, S.P.: Software agents. Commun. ACM **37**(7), 48–53 (1994)
2. Wooldridge, M., Jennings, N.R.: Intelligent agents: theory and practice. Knowl. Eng. Rev. **10**(2), 115–152 (1995)
3. Russell, S.J., Norvig, P.: Artificial Intelligence: A Modern Approach, 3rd edn. Pearson Education Inc., Prentice Hall (2010)
4. Bellifemine, F., Caire, G., Greenwood, D.: Developing Multi-agent Systems with JADE. Wiley, Chichester (2007)

5. Wooldridge, M.: Agent-based computing. Interoper. Commun. Netw. **1**, 71–98 (1998)
6. Niazi, M., Hussain, A.: Agent-based computing from multi-agent systems to agent-based models: a visual survey. Scientometrics **89**(2), 479–499 (2011)
7. Weiss, G. (ed.): Multiagent Systems: A Modern Approach to Distributed Artificial Intelligence. The MIT Press, Cambridge, MA (1999)
8. Wooldridge, M.: An Introduction to Multiagent Systems, 2nd edn. Wiley (2009)
9. Madejski, J.: Survey of the agent-based approach to intelligent manufacturing. J. Achiev. Mater. Manuf. Eng. **21**(1), 67–70 (2007)
10. Balaji, P.G., Srinivasan, D.: An introduction to multi-agent systems. In: Innovations in Multi-agent Systems and Applications — 1, Studies in Computational Intelligence, vol. 310, pp. 1–27 (2010)
11. Vlassis N.: A concise introduction to multiagent systems and distributed artificial intelligence. In: Synthesis Lectures on Artificial Intelligence and Machine Learning. Morgan & Claypool (2007)
12. Burke, E.K., Graham Kendall, G.: Search Methodologies: Introductory Tutorials in Optimization and Decision Support Techniques. Springer, US (2014)
13. Jennings, N.R., Wooldridge, M.: Applications of intelligent agents. In: Jennings, N.R., Wooldridge, M.J. (eds.) Agent Technology: Foundations, Applications, and Markets, pp. 3–28. Springer, Berlin (1998)
14. Jennings, N.: The archon system and its applications. In: Proceedings of the 2nd International Working Conference on Cooperating Knowledge Based Systems (CKBS-94), pp. 13–29. Dake Centre, University of Keele, UK (1994)
15. Albert, M., Laengle, T., Woern, H., Capobianco, M., Brighenti, A.: Multi-agent systems for industrial diagnostics. In: Proceedings of 5th IFAC Symposium on Fault Detection, Supervision and Safety of Technical Processes, pp. 483–488, Washington, DC (2003)
16. Neagu, N., Dorer, K., Greenwood, D., Calisti, M.: LS/ATN: reporting on a successful agent-based solution for transport logistics optimization. In: Proceedings of the IEEE 2006 Workshop on Distributed Intelligent Systems (WDIS06), Prague (2006)
17. Greenwood, D., Vitaglione, G., Keller, L., Calisti, M.: Service level agreement management with adaptive coordination. In: Proceedings of the International Conference on Networking and Services (ICNS06), Silicon Valley, USA (2006)
18. Johnson, P.G., Balke, T., Kotthoff, L.: Integrating optimisation and agent-based modelling. In: ECMS — Proceedings 28th European Conference on Modelling and Simulation, pp. 775–781. Digitaldruck Pirrot GmbH, Germany (2014)
19. Parunak, H.V.D., Kindrick, J., Irish, B.W.: A conservative domain for neural connectivity and propagation. In: Proceedings of Proceedings of the 6th National Conference on Artificial Intelligence (AAAI'87). Distributed artificial intelligence, pp. 307–311. Pitman, London (1987)
20. Sirikijpanichkul, A., van Dam, K.H., Ferreira, L., Lukszo, Z.: Optimizing the location of intermodal freight hubs: an overview of agent based modelling approach. J. Transp. Syst. Eng. Inf. Technol. **7**(4), 71–81 (2007)
21. Ouelhadj, D., Petrovic, S.: A survey of dynamic scheduling in manufacturing systems. J. Sched. **12**(4), 417–431 (2008)
22. Böcker, J., Lind, J., Zirkler, B.: Using a multi-agent approach to optimise the train coupling and sharing system. Eur. J. Oper. Res. **131**(2), 242–252 (2010)
23. Liang, W.Y., Huang, C.C.: Agent-based demand forecast in multi-echelon supply chain. Decis. Support Syst. **42**(1), 390–407 (2006)
24. Barbucha, D., Jędrzejowicz, P.: An agent-based approach to vehicle routing problem. In. J. Appl. Math. Comput. Sci. **4**(2), 538–543 (2007)
25. Polyakovsky, S., M'Hallah, R.: An agent-based approach to the two dimensional guillotine bin packing problem. Eur. J. Oper. Res. **192**(31), 767–781 (2009)
26. Xie, X.F., Liu, J.: Multiagent optimization system for solving the traveling salesman problem (TSP). IEEE Trans. Syst. Man Cybern. Part B Cybern. **39**(2), 489–502 (2009)
27. Barbati, M., Bruno, G., Genovese, A.: Applications of agent-based models for optimization problems: a literature review. Expert Syst. Appl. **39**, 6020–6028 (2012)

28. Knotts, G., Dror, M., Hartman, B.C.: Agent-based project scheduling. IIE Trans. **32**(5), 387–401 (2000)
29. Chen, Y.M., Wang, S.C.: Framework of agent-based intelligence system with two stage decision-making process for distributed dynamic scheduling. Appl. Soft Comput. **7**(1), 229–245 (2007)
30. Xiang, W., Lee, H.P.: Ant colony intelligence in multi-agent dynamic manufacturing scheduling. Eng. Appl. Artif. Intell. **21**(1), 73–85 (2008)
31. Ramos, C.: An architecture and a negotiation protocol for the dynamic scheduling of manufacturing systems. In: Proceedings of IEEE International Conference on Robotics and Automation, pp. 8–13 (1994)
32. Davidsson, P., Holmgren, J., Persson, J.A.: On the integration of agent-based and mathematical optimization techniques. Lect. Notes Artif. Intell. **4496**, 1–10 (2007)
33. Chen, R.S., Tu, M.A.: Development of an agent-based system for manufacturing control and coordination with ontology and RFID technology. Expert Syst. Appl. **36**(4), 7581–7593 (2009)
34. Persson, J.A., Davidsson, P., Johansson, S.J., Wernstedt, F.: Combining agent-based approaches and classical optimization techniques. In: Proceedings of the Third European Workshop on Multi-Agent Systems (EUMAS 2005), pp. 260–269 (2005)
35. Talukdar, S., Baerentzen, L., Gove, A., De Souza, P.: Asynchronous Teams: Co-operation Schemes for Autonomous, Computer-Based Agents. Technical Report EDRC 18-59-96, Carnegie Mellon University, Pittsburgh (1996)
36. Talukdar, S., Baerentzen, L., Gove, A., de Souza, P.: Asynchronous teams: cooperation schemes for autonomous agents. J. Heuristics **4**(4), 295–332 (1998)
37. Barbucha, D., Czarnowski, I., Jędrzejowicz, P., Ratajczak-Ropel, E., Wierzbowska, I.: e-JABAT — An implementation of the web-based A-Team. In: Nguyen, N.T., Jain, L.C. (eds.) Intelligence Agents in the Evolution of Web and Applications. Studies in Computational Intelligence 167, 57–86 (2009)
38. Jędrzejowicz, P., Wierzbowska, I.: JADE-based A-Team environment. In: Computational Science — ICCS. Lecture Notes in Computer Science, vol. 3993, pp. 719–726 (2006)
39. Talukdar, S.N., de Souza, P.: Scale efficient organizations. In: IEEE International Conference on Systems, Man, and Cybernetics, Chicago, pp. 1458–1463 (1992)
40. Talukdar, S., Murthy, S., Akkiraju, R.: Asynchronous teams. In: Handbook of Metaheuristics. International Series in Operations Research & Management Science, vol. 57, pp. 537–556 (2003)
41. Correa, R., Gomes, F.C., Oliveira, C., Pardalos, P.M.: A parallel implementation of an asynchronous team to the point-to-point connection problem. Parallel Comput. **29**, 447–466 (2003)
42. Zhu, Q.: Topologies of agents interactions in knowledge intensive multi-agent systems for networked information services. Adv. Eng. Inform. **20**, 31–45 (2006)
43. Jędrzejowicz, P.: A-Teams and their applications. In: Nguyen, N.T., Kowalczyk, R., Chen, S.-M. (eds.) ICCCI 2009. Lecture Notes in Computer Science(LNAI), vol. 5796, pp. 36–50 (2009)
44. Carle, M.A., Martel, A., Zufferey, N.: Collaborative Agent Teams (CAT) for Distributed Multi-Dimensional Optimization. CIRRELT, CIRRELT-2012-43, (2012)
45. Talukdar, S.N.: Collaboration rules for autonomous software agents. Decis. Support Syst. **24**, 269–278 (1999)
46. Rachlin, J., Goodwin, R., Murthy, S., Akkiraju, R., Wu, F., Kumaran, S., Das, R.: A-Teams: an agent architecture for optimization and decision-support. In: Papadimitriou, C., Singh, M.P., Müller, J.P. (eds.) ATAL 1998. Lecture Notes in Artificial Intelligence, vol. 1555, pp. 261–276 (1999)
47. Barbucha, D., Czarnowski, I., Jędrzejowicz, P., Ratajczak, E., Wierzbowska, I.: JADE-Based A-Team as a tool for implementing population-based algorithms. In: Chen, Y., Abraham, A. (eds.) Intelligent Systems Design and Applications, Jinan Shandong, China, pp. 144–149. IEEE, Los Alamitos (2006)
48. Talukdar, S.N., Ramesh, V.C.: A multi-agent technique for contingency constrained optimal power flows. IEEE Trans. Power Syst. **9**(2), 855–861 (1994)

49. Avila-Abascal, P., Talukdar, S.N.: Cooperative algorithms and abductive causal networks for the automatic generation of intelligent substation alarm processors. In: Proceedings of ISCAS'96 (1996)
50. Kao, J.H., Hemmerle, J.S., Prinz, F.B.: Collision avoidance using asynchronous teams. In: 1996 IEEE International Conference on Robotics and Automation, vol. 2, pp. 1093–1100. OMNI Press, USA (1996)
51. Murthy, S., Rachlin, J., Akkiraju, R., Wu, F.: Agent-based cooperative scheduling. In: Charniak, E.C. (ed.) Constraints and Agents, AAAI Technical Report WS-97-05, pp. 112–117 (1997)
52. Blum, J., Eskandarian, A.: Enhancing intelligent agent collaboration for flow optimization of railroad traffic. Transp. Res. **36**(10), 919–930 (2002)
53. Chen, S.Y., Talukdar, S. N., Sadeh N. M.: Job-Shop-Scheduling by a team of asynchronous agents. In: IJCAI-93 Workshop on Knowledge-Based Production, Scheduling and Control, Chambery, France (1993)
54. Aydin, M.E., Fogarty, T.C.: Teams of autonomous agents for job-shop scheduling problems: an experimental study. J. Intell. Manuf. **15**, 455–462 (2004)
55. Aydin, M.: Metaheuristic agent teams for job shop scheduling problems. In: Holonic and Multi-Agent Systems for Manufacturing. Lecture Notes in Computer Science, vol. 4659, pp. 185–194 (2007)
56. Rachlin, J., Wu, F., Murthy, S., Talukdar, S., Sturzenbecker, M., Akkiraju, R., Fuhrer, R., Aggarwal, A., Yeh, J., Henry, R., Jayaraman, R.: ForestView: a system for integrated scheduling in complex manufacturing domains. IBM Report (1996)
57. Lee, H., Murthy, S., Haider, W., Morse, D.: Primary production scheduling at steel making industries, IBM Report (1995)
58. Tsen, C.K.: Solving train scheduling problems using A-Teams. Ph.D. dissertation, Electrical and Computer Engineering Department, CMU, Pittsburgh, PA (1995)
59. Rabak, C.S., Sichman, J.S.: Using A-Teams to optimize automatic insertion of electronic components. Adv. Eng. Inform. **17**, 95–106 (2003)
60. Czarnowski, I., Jędrzejowicz, P.: Agent-based NON-distributed and distributed clustering. In: Perner, P. (ed.) Machine Learning and Dara Mining in Pattern Recognition. Lecture Notes in Artificial Intelligence, vol. 5632, pp. 347–360. Springer, Berlin, Heidelberg (2009)
61. Jędrzejowicz, P., Wierzbowska, I. Parallel cooperating A-Teams solving instances of the euclidean planar traveling salesman problem. In: J. O'Shea et al. (eds.) Agent and Multi Agent Systems: Technologies and Applications. Lecture Notes in Artificial Intelligence, vol. 6682, pp. 456–465 (2011)
62. Carle, M.A., Martel, A., Zufferey, N.: The CAT metaheuristic for the solution of multi-period activity-based supply chain network design problems. Int. J. Prod. Econ. **139**(2), 664–677 (2012)
63. Barbucha, D.: Experimental Study of the Population Parameters Settings in Cooperative Multi-agent System Solving Instances of the VRP. In: Transactions on Computational Collective Intelligence IX. Lecture Notes in Computer Science, vol. 7770, pp. 1–28 (2013)
64. Barbucha, D.: A cooperative population learning algorithm for vehicle routing problem with time windows. Neurocomputing **146**, 210–229 (2014)
65. Barbucha, D.: Team of A-Teams approach for vehicle routing problem with time windows. In: Terrazas, G., Otero, F., Masegosa, A. (eds.) Nature Inspired Cooperative Strategies for Optimization (NICSO 2013), vol. 512, pp. 273–286. Springer International Publishing (2014)
66. Barbucha, D., Czarnowski, I., Jędrzejowicz, P., Ratajczak-Ropel, E., Wierzbowska, I.: Influence of the working strategy on A-Team performance. In: Szczerbicki, E., Nguyen, N.T. (eds.) Smart Information and Knowledge Management. Studies in Computational Intelligence, vol. 260, pp. 83–102. Springer, Heidelberg (2010)
67. Bellifemine, F., Caire, G., Poggi, A., Rimassa, G.: JADE. A White Paper, Exp. **3**(3), 6–20 (2003)
68. JADE (Java Agent DEvelopment framework). http://jade.tilab.com/

69. Barbucha, D., I. Czarnowski, P. Jędrzejowicz, E. Ratajczak-Ropel, I. Wierzbowska: JABAT — an implementation of the A-Team concept. In: Proceedings of the International Multiconference on Computer Science and Information Technology, vol. 1, pp. 235–241. Polskie Towarzystwo Informatyczne, Wisła (2006)
70. Barbucha, D., Czarnowski, I., Jędrzejowicz, P., Ratajczak-Ropel, E., Wierzbowska, I.: Parallel cooperating A-Teams, In: P.Jędrzejowicz et al. (eds.) Computational Collective Intelligence. Technologies and Applications. Lecture Notes in Artificial Intelligencc, vol. 6923, pp. 322–331. Springer, Heidelberg (2011)
71. Jędrzejowicz, P., Ratajczak-Ropel, E.: A-Team for solving the resource availability cost problem. In: Nguyen, N.T., Hoang, K., Jędrzejowicz, P. (eds.) Computational Collective Intelligence Technologies and Applications. Lecture Notes in Artificial Intelligence, vol. 7654, pp. 443–452 (2012)
72. Barbucha, D., Czarnowski, I., Jędrzejowicz, P., Ratajczak-Ropel, E., Wierzbowska, I.: Team of A-Teams — A study of the cooperation between program agents solving difficult optimization problems, Agent-Based Optimization. In: Czarnowski, I., Jędrzejowicz, P., Kacprzyk, J. (eds) Studies in Computational Intelligence, vol. 456, pp. 123–142. Springer, Heidelberg (2013)
73. Jędrzejowicz, P., Ratajczak-Ropel, E.: Reinforcement learning strategy for solving the resource-constrained project scheduling problem by a team of A-Teams. In: Nguyen, N.T., Attachoo, B., Trawiński, B., Somboonviwat, K. (eds.) Intelligent Information and Database Systems. Lecture Notes in Artificial Intelligence, vol. 8398, pp. 197–206 (2014)
74. Ren, H., Wang, Y.: A survey of multi-agent methods for solving resource constrained project scheduling problems. In: Proceedings of International Conference on Management and Service Science, vol. 2011, pp. 1–4 (2011)
75. Jędrzejowicz, P., Ratajczak-Ropel, E.: Agent-Based Approach to Solving the Resource Constrained Project Scheduling Problem. Lecture Notes in Computer Science, vol. 4431, pp. 480–487 (2007)
76. Jędrzejowicz, P., Ratajczak-Ropel, E.: New generation A-Team for solving the resource constrained project scheduling. In: Proceedings of the Eleventh International Workshop on Project Management and Scheduling, pp. 156–159. Istanbul (2008)
77. Jędrzejowicz, P., Ratajczak-Ropel, E.: Solving the RCPSP/max problem by the team of agents. In: Hakansson, A., et al. (eds.) Agent and Multi-Agent Systems: Technologies an Applications. Lecture Notes in Artificial Intelligence, vol. 5559, pp. 734–743 (2009)
78. Jędrzejowicz, P., Ratajczak-Ropel, E.: Team of A-Teams for solving the resource-constrained project scheduling problem. In: Grana, M., Toro, C., Posada, J., Howlett, R., Lakhmi, C.J. (eds.) Advances in Knowledge Based and Intelligent Information and Engineering Systems. Frontiers in Artificial Intelligence and Applications, vol. 243, pp. 1201–1210, (2012)
79. Jędrzejowicz, P., Ratajczak-Ropel, E.: Reinforcement learning strategies for A-Team solving the resource-constrained project scheduling problem. Neurocomputing **146**, 301–307 (2014)
80. Jędrzejowicz, P., Ratajczak-Ropel, E.: Reinforcement Learning Strategy for Solving the MRCPSP by a Team of Agents. In: Neves-Silva, R., Jain, L.C., Howlett, R.J. (eds.) Intelligent Decision Technologies, Proceedings of the 7th KES International Conference on Intelligent Decision Technologies (KES-IDT 2015), pp. 537–548. Springer International Publishing, Switzerland (2015)
81. Jędrzejowicz, P., Ratajczak-Ropel, E.: PLA Based Strategy for Solving RCPSP by a Team of Agents. J. Univ. Comput. Sci. **22**(6), 856–873 (2016)
82. Jędrzejowicz P., Ratajczak-Ropel E.: Dynamic cooperative interaction strategy for solving RCPSP by a team of agents. In: Nguyen N.T., Manolopoulos, Y., Iliadis, L., Trawiński, B. (eds.) Computational Collective Intelligence. Lecture Notes in Artificial Intelligence, vol. 9875, pp. 454–463 (2016)

Chapter 3
Project Scheduling Models

Project scheduling problems (PSP) could be defined as allocating scarce resources over time to perform a given set of activities. The resources are arbitrary although constituting a necessary means which activities compete for. Notion of the activity can have a variety of interpretations. Thus defined project scheduling problems appear in a large spectrum of real-world situations, and, in consequence, have been intensively studied by specialists in management science, operations research and computer science.

In this chapter the brief history of the deterministic project scheduling models (formulations) is presented in Sect. 3.1. Section 3.2 review basic models for the Resource-Constrained Project Scheduling Problem (RCPSP) as well as their classification. Finally, in Sect. 3.3 some special cases of the RCPSP are mentioned as a short review of its big family. Examples of most commonly used objective functions are presented in Sect. 3.4.

3.1 Historical Review

The first models and methods for deterministic project scheduling date back to the 1950s when the well known network-based models like critical path method or project evaluation and review technique were formulated and developed. Critical Path Method (CPM) [1, 2] is based on mathematical model which determines the sequence of project activities for projects with ordinary precedence constraints. In Metra Potential Method (MPM) [3] generalized precedence constraints are considered. Project Evaluation and Review Technique (PERT) [4] focuses on creating and controlling project schedules in stochastic environments using deterministic or probabilistic activity processing times. Although the network-based models have proved useful in handling scheduling for various projects, they simplified the problems by assuming that the availability of resources is not limited, which is

E. Ratajczak-Ropel and A. Skakovski, *Population-Based Approaches to the Resource-Constrained and Discrete-Continuous Scheduling*,
Studies in Systems, Decision and Control 108, DOI 10.1007/978-3-319-62893-6_3

unrealistic in majority of practical situations. Beginning in the late 1960s, the models were extended by additionally considering scarcity of resources, see for example [5, 6]. The impact of limited resources on project characteristics started to be taken into account in such models as CPM/MCX (Minimum Cost eXpediting) or CPM/resources. Graphical Evaluation and Review Technique (GERT) [7], in turn, allows loops between activities and additionally takes probabilistic precedence relations into account. These problems and models are usually referred to as the resource-constrained.

Since then, interest and research efforts in the field of resource-constrained project scheduling have increased, and many new models and methods have been developed. Overviews of the advances in models and solution methods are given in the survey papers of Icmeli et al. [8], Elmaghraby [9], Özdamar and Ulusoy [10], Herroelen et al. [11], Brucker et al. [12], Kölisch and Padman [13], Dorndorf [14], Artigues et al. [15] or Węglarz et al. [16].

3.2 Basic Models and Classifications Review

Most of the early studies on RCPSP relate to mathematical programming formulation and its relaxation used in B&B (branch-and-bound) approaches [17, 18]. However, solving the linear programming relaxation of any model of the RCPSP is too time consuming, or just impossible, in practical applications [15]. The RCPSP can be formulated as a mathematical programming model in several ways, depending on the definition of the decision variables and constraints construction.

The main mathematical programming formulations and their relaxations proposed for the classical RCPSP can be divided into two main classes [15]:

- sequence/order-based models,
- time-indexed/discrete-time integer linear programming models.

The first class concentrates on ordering, the sequence activities are processed one after another. The second class concentrates on assigning resources to activities at each point of time.

In the sequence based mathematical programming models scheduling can be viewed as determining a sequence of activities satisfying the precedence constraints, and then fixing the processing times of activities following such sequence and respecting other possible constraints. In the time-indexed models problem formulations are based on time discretization which naturally describe the usage of the resources and the processing of the activities over time. In these models the fixed planning horizon, denoted by H, is required. It means that all activities have to be completed by time H.

In recent years different classifications for the RCPSP have been proposed and used, and new models have appeared. Reviews and propositions of the new models can be found for example in the papers of Koné et al. [19, 20], Artigues [21], and Kopanos et al. [22].

The RCPSP models are formulated as:

- Binary Integer Programming (BIP),
- Mixed Integer Programming (MIP),
- Mixed Integer Linear Programming (MILP).

In binary formulation, each variable can only take on thc value of 0 or 1. In the RCPSP binary variables are usually used to indicate whether one activity is processed before the other or whether an activity finishes (or starts) at fixed time point or not. In the MIP formulation some of the variables are real-valued and some of the variables are integer-valued. When the objective function and constraints are all linear, it is called MILP.

RCPSP models could be also classified as:

- Discrete-Time (DT),
- Continuous-Time (CT).

In DT models time-indexed binary decision variables are used. Hence, schedule events can only take place at a certain predefined time points. In such models the time horizon is divided in uniform time intervals. In CT models for the RCPSP, binary variables indicate the processing sequence between pairs of activities, hence rely on precedence-based decision variables. In such models schedule events can occur at any time in the time horizon of interest. Time-indexed CT models can be derived, if a variable time grid is used.

Several time-indexed DT BIP models were proposed for the RCPSP. One of the first such model, where one linear constraint for each time period is formulated to avoid resource conflicts, was proposed by Pritsker et al. [17] (see Sect. 4.1). It is called basic discrete-time (DT) model. Very similar formulation, called disaggregated discrete-time (DDT) formulation was proposed by Christofides et al. [23]. The mainly difference is in the precedence constraints formulation.

Similar model, based on the notion of the feasible set of activities, was proposed by Mingozzi et al. [24]. Additionally, they derived lower bounds from this formulation by relaxing the non-preemption and the precedence constraints to disjunctions and time-windows. The relaxation is applied to calculating lower bounds for different scheduling problems.

Klein [25] adopted to the classical RCPSP a DT BIP formulation proposed for preemptive version of the RCPSP by Kaplan [26], where a single type of binary variables that specifies whether activity is active over time period or not was introduced, as well as a simpler definition of the resource-constraints. Klein [25] proposed two additional models for the RCPSP. The first is based on the definition of binary variables which specify if an activity starts at the beginning of fixed time period or earlier. In the second model a different range for the above mentioned binary variables and a new binary variables, that denote whether an activity is completed at the end of fixed time period or earlier, were introduced.

Artigues et al. [27] proposed a formulation called Flow-based Continuous-Time (FCT) model. It is based on the notation of resource flow, where three types of variables are used: starting time variables for each activity, sequential binary variables indicating whether one activity is processed before the other, and flow variables denoting the quantity of resource that is transferred between activities.

The hybrid between a sequence-based model and Mingozzi's relaxation mathematical programming model was proposed by Carlier and Néron [28, 29].

Several CT MIP models were proposed for the RCPSP. One of the first, based on minimal forbidden (resource incompatible) sets and sequence of activities, was proposed by Alvarez-Valdez and Tamarit [30]. These sets contain activities that have no precedence relations between them and cannot be executed simultaneously due to resource availability. It is a natural extension of the linear ordering approach for disjunctive scheduling. In this model binary variables define the sequencing of activities, and integer variables represent the starting time of each activity.

Koné et al. in [19] proposed two event-based CT MIP models based on the concept of event: the Start/End and On/Off formulation. In these models the decision variables are indexed using event points (instead of time points) that correspond to the start or end times of activities. The scheduling horizon is subdivided into intervals of variable length. An event means the beginning of the interval when the activity starts. Each activity starts at a unique event and the starting time of the activity equals the starting time of the assigned event. In the Start/End Event-based Model (SEE) two sets of binary variables and two types of the continuous variables are used. Binary variables describe the activity starts and ends at any event point. Continuous variables represent the date of event and the quantity of resources required immediately after the event. In the On/Off Event-based Model (OOE) only one type of binary variable per event, and one type of continuous variable are used. The binary variable denotes whether the activity starts processing at the event point or it is still being processed immediately after this event. The continuous variable represents, as in SEE, the date of event. In this model, the number of events is exactly equal to the number of activities.

Bianco and Caramia [31] proposed a DT MIP formulation based on the definition of one type of the continuous variable that represents the percentage of the activity that is executed at each time interval. Two types of binary variables represent whether the activity starts and ends at the time interval.

Recently, Kyriakidis et al. [32] proposed a time-indexed CT MIP formulation for single- and multi-mode RCPSP. It is based on the Resource-Task Network (RTN) representation. RTN is a network representation technique, used in process scheduling problems, based on the continuous time models.

Moreover, Kopanos et al. [22] proposed two DT MILP models and two CT MILP models. The DT models are based on the definition of binary variables that describe the processing state of every activity between two consecutive time points, The continuous-time models are based on the concept of overlapping of activities, and the definition of a number of newly introduced sets.

The other classical representation, as well as method of solving the RCPSP is based on the constraint programming. The classical RCPSP was formulated as a constraint-based scheduling approach, using the Optimization Programming Language (OPL), by Van Hentenryck [33].

3.3 Generalizations and Special Cases of the RCPSP

While the RCPSP is already a powerful model, it does not cover all situations that occur in practice. On the other hand, RCPSP model is commonly used in a variety of engineering fields. Hence, many scheduling problems can be formulated as the special case of the RCPSP, such as job-shop scheduling, flow-shop scheduling, open-shop scheduling and project scheduling [34]. Therefore, many more general project scheduling models have been developed together with extensions and variants of the RCPSP such as well known MRCPSP, MSPSP, RCPSP/max, MRCPSP/max, or MRCMPSP.

The MRCPSP (Multi-mode Resource-Constrained Project Scheduling Problem) is a generalization of the RCPSP where several modes of execution may exist for any activity of the project. The problem is considered in Sect. 5. In this model also the renewable and non-renewable resources are considered.

An example of special case of the MRCPSP is the Discrete Time/Resource Trade-off Problem (DTRTP) where only one single renewable resource with fixed capacity is given (e.g., modeling available staff).

The MSPSP (Multi-Skill Project Scheduling Problem) was proposed by Néron and Baptista [35] to model project scheduling problems in IT companies. In this extension, resources correspond to persons that master a subset of skills required by the different activities of the project.

The RCPSP/max (RCPSP with minimum and maximum time lags), known also in the literature as the RCPSP with time windows or RCPSP with Generalized Precedence Relations (RCPSP-GPR), involves generalized precedence relations (temporal constraints) of the start-to-start type where the minimum and maximum time lags between activities are considered. The MRCPSP/max is generalization of the RCPSP/max, MRCPSP and RCPSP.

The MRCMPSP (Multi-mode Resource-Constrained Multi-Project Scheduling Problem) is a generalization of RCPSP and MRCPSP where additionally multiple projects are considered [36]. The MRCMPSP consists of a number of projects, defined as collections of activities performed in one of several ways under given precedence relationships and limited amounts of various types of resources.

In the RCPSPVRL (Resource-Constrained Project Scheduling Problem with Varying Resource Levels) resources of limited availability but varying predetermined levels are used. In this problem project activities are not assumed to have constant resources throughout the entire project duration. In RCPSPVRL, resources are not constant throughout the entire project duration. The total project duration is divided into different time periods. Within a particular time period quantity of resources can vary in various time periods.

3.4 Objective Functions

The objective of the basic PSP model is time minimization, however the resource, cost or quality related objectives are also commonly used, especially in practical applications. For example: the resource-based objective functions are considered in the area of resource investment problems (RIP), resource leveling problems (RLP) and time/resource trade-off problem (TRTP); the cost-based objective functions in the area of time/cost trade-off problems (TCTP) and payment scheduling problems (PSP).

In this part of the book the total length of the schedule minimization is considered, which is the most commonly used objective function in the project scheduling. The total length of the schedule is also called project duration, project complexion time, or makespan.

The other commonly used objective functions include, for example:

- total flow time,
- weighted (total) flow time,
- maximum lateness,
- total tardiness,
- total weighted tardiness,
- number of late activities,
- weighted number of late activities,
- net present value (NPV).

In case of optimizing one criterion only the single objective optimization problems are considered. If there are more than one criterion which must be treated simultaneously, the multiple objective optimization problems must be taken into account. The models for multiple objective project scheduling were considered, among others, by Davis [37] and Ishii [38].

References

1. Kelley, J.E., Jr., Walker, M.R.: Critical path planning and scheduling. In: Proceedings of the Eastern Joint Computer Conference, pp. 160–173, Boston, MA (1959)
2. Kelly, J.: Critical path planning and scheduling: mathematical basis. Oper. Res. **9**, 296–320 (1961)
3. Roy, B.: Graphes et Ordonnancement. Revue Française de Recherche Opérationelle, pp. 323–333 (1962)
4. Malcolm, D., Roseboom, J., Clark, C., Fazar, W.: Applications of a technique for research and development program evaluation. Oper. Res. **7**, 646–669 (1959)
5. Wiest, J.D.: Some properties of schedules for large projects with limited resources. Oper. Res. **12**, 395–418 (1964)
6. Davis, E.W.: Resource allocation in project network models — a survey. J. Ind. Eng. **17**(4), 177–188 (1966)
7. Pritsker, A., Happ, W.W.: GERT: Graphical evaluation and review technique — Part I: Fundamentals. J. Ind. Eng. **17**, 267–274 (1966)

8. Icmeli, O., Erenguc, S.S., Zappe, C.J.: Project scheduling problems: a survey. Int. J. Oper. Prod. Manag. **13**(11), 80–91 (1993)
9. Elmaghraby, S.E.: Activity nets: a guided tour through some recent developments. Eur. J. Oper. Res. **82**(3), 383–408 (1995)
10. Özdamar, L., Ulusoy, G.: A survey on the resource-constrained project scheduling problem. IIE Trans. **27**(5), 574–586 (1995)
11. Herroelen, W.S., De Reyck, B., Demeulemeester, E.L.: Resource-constrained project scheduling: a survey of recent developments. Comput. Oper. Res. **25**(4), 279–302 (1998)
12. Brucker, P., Drexl, A., Möhring, R., Neumann, K., Pesch, E.: Resource-constrained project scheduling: notation, classification, models, and methods. Eur. J. Oper. Res. **112**, 3–41 (1999)
13. Kölisch, R., Padman, R.: An integrated survey of deterministic project scheduling. OMEGA Int. J. Manag. Sci. **29**(3), 249–272 (2001)
14. Dorndorf., U.: Project Scheduling with Time Windows: From Theory to Applications. Physica-Verlag (2002)
15. Artigues, C., Demassey, S., Neron, E.: Resource-Constrained Project Scheduling: Models, Algorithms, Extensions and Applications. In: Control Systems, Robotics and Manufacturing Series. ISTE/Wiley, (2008)
16. Węglarz, J., Józefowska, J., Mika, M., Waligóra, G.: Project scheduling with finite or infinite number of activity processing modes — a survey. Eur. J. Oper. Res. **208**, 177–205 (2011)
17. Pritsker, A.A.B., Watters, L.J., Wolfe, P.M.: Multi-project scheduling with limited resources: a zero-one programming approach. Manag. Sci. **16**(1), 93–108 (1969)
18. Balas, E.: Project scheduling with resource constraints. In: Beale, E.M.L. (ed.) Applications of Mathematical Programming Techniques, pp. 187–200. American Elsevier, New York (1970)
19. Koné, O., Artigues, C., Lopez, P., Mongeau, M.: Event-based MILP models for resource-constrained project scheduling problems. Comput. Oper. Res. **38**(1), 3–13 (2011)
20. Koné, O., Artigues, C., Lopez, P., Mongeau, M.: Comparison of mixed integer linear programming models for the resource-constrained project scheduling problem with consumption and production of resources. Flex. Serv. Manuf. J. **25**(1–2), 24–47 (2013)
21. Artigues, C.: A note on time-indexed formulations for the resource-constrained project scheduling problem. Rapport LAAS n 13206 (2013)
22. Kopanos, G.M., Kyriakidis, T.S., Georgiadis, M.C.: New continuous-time and discrete-time mathematical formulations for resource-constrained project scheduling problems. Comput. Chem. Eng. **68**, 96–106 (2014)
23. Christofides, N., Alvarez-Valdes, R., Tamarit, J.M.: Project scheduling with resource constraints: a branch and bound approach. Eur. J. Oper. Res. **29**, 262–273 (1987)
24. Mingozzi, A., Maniezzo, V., Ricciardelli, S., Bianco, L.: An exact algorithm for project scheduling with resource constraints based on a new mathematical formulation. Manag. Sci. **44**(5), 714–729 (1998)
25. Klein, R.: Scheduling of Resource-Constrained Projects. Kluwer Academic Publishers, Boston (2000)
26. Kaplan, L.A.: Resource-constrained project scheduling with preemption of jobs. Ph.D. thesis, University of Michigan (1988)
27. Artigues, C., Michelon, P., Reusser, S.: Insertion techniques for static and dynamic resource-constrained project scheduling. Eur. J. Oper. Res. **149**, 249–267 (2003)
28. Carlier, J., Néron, E.: On linear lower bounds for the resource constrained project scheduling problem. Eur. J. Oper. Res. **149**, 314–324 (2003)
29. Carlier, J., Néron, E.: Computing redundant resources for the resource constrained project scheduling problem. Eur. J. Oper. Res. **176**, 1452–1463 (2007)
30. Alvarez-Valdés, R., Tamarit, J.M.: The project scheduling polyhedron: dimension, facets and lifting theorems. Eur. J. Oper. Res. **67**(2), 204–220 (1993)
31. Bianco, L., Caramia, M.: A new formulation for the project scheduling problem under limited resources. Flex. Serv. Manuf. J. **25**(1–2), 6–24 (2013)
32. Kyriakidis, T.S., Kopanos, G.M., Georgiadis, M.C.: MILP formulations for single- and multi-mode resource-constrained project scheduling problems. Comput. Chem. Eng. **36**, 369–385 (2012)

33. Hentenryck Van, P.: The OPL Optimization Programming Language. MIT Press (1999)
34. Leung, J.Y.: Handbook of Scheduling: Algorithms, Models, and Performance Analysis. In: Computer and Information Science Series. A CRC Press Company, Chapman & Hall/CRC (2004)
35. Néron, E., Baptista, D.: Heuristics for the multi-skill project scheduling problem. In: International Symposium on Combinatorial Optimization (CO'2002). France, Paris (2002)
36. Tseng, C.-C.: Two heuristic algorithms for a multi-mode resource-constrained multi-project scheduling problem. J. Sci. Eng. Technol. **4**(2), 63–74 (2008)
37. Davis, K.R.: Resource constrained project scheduling with multiple objectives: a decision support approach. Comput. Oper. Res. **19**(7), 657–669 (1992)
38. Ishii, H.: Multiobjective Scheduling Problems. Lect. Notes Econ. Math. Syst. **405**, 386–391 (1994)

Chapter 4
Resource-Constrained Project Scheduling

The first part of this chapter presents Resource-Constrained Project Scheduling Problem (RCPSP) formulations and notations (Sect. 4.1). It also provides an overview of the best methods proposed so far for solving this problem, including a set of relevant bibliographic references in Sect. 4.2.

The second part of this chapter in Sects. 4.3 and 4.4 provides an overview of agent-based approaches suggested for solving RCPSP. This part includes methods, based on the A-Team approach, proposed by the author. Finally, in Sect. 4.4.7 comparison of results is presented and some concluding remarks are formulated.

4.1 Problem Formulation

The Resource-Constrained Project Scheduling Problem (RCPSP) is a classical discrete problem, i.e. the planning horizon is divided into a discrete number of time periods with the discrete activity durations and resource units.

In the RCPSP formulation an abstract project is considered. It consists of the set of n activities, where each activity has to be processed without interruption to complete the project. The additional dummy activities 0 and $n+1$ represent the beginning and the end of the project. The duration of an activity $j, j = 0, \ldots, n+1$ is denoted by d_j where $d_0 = d_{n+1} = 0$. There is the set R of r renewable resource types. The availability of each resource type k in each time period is r_k units, $k = 1, \ldots, r$. Each activity j requires r_{jk} units of resource k during each period of its duration, where $r_{0k} = r_{n+1k} = 0, k = 1, \ldots, r$. All parameters are non-negative integers. There are precedence relations of the finish-start type with a zero parameter value (i.e. $FS = 0$) defined between the activities. In other words activity i precedes activity j if j cannot start until i has been completed.

The structure of a project can be represented by the Activity On Node (AON) network $G = (V, E)$, where V denotes the set of activities and E is the set of precedence relationships between these activities. S_j (P_j) is the set of successors (predecessors)

E. Ratajczak-Ropel and A. Skakovski, *Population-Based Approaches to the Resource-Constrained and Discrete-Continuous Scheduling*,
Studies in Systems, Decision and Control 108, DOI 10.1007/978-3-319-62893-6_4

of activity j, $j = 0, \ldots, n+1$. It is further assumed that $0 \in P_j$, $j = 1, \ldots, n+1$, and $n+1 \in S_j$, $j = 0, \ldots, n$.

The objective is to find a schedule S of activities starting times $[s_0, \ldots, s_{n+1}]$, where $s_0 = 0$ and the precedence and resource-constraints are satisfied, such that the project complexion time (schedule duration, makespan) $CT(S) = s_{n+1}$ is minimized.

The RCPSP may be formulated as the mathematical, or more precisely the integer programming problem. The formulation was introduced by Pritsker et al. in [1]. The $0-1$ decision variable $x_{jt} = 1$ if activity j is assigned a completion time at the end of period t, otherwise, $x_{jt} = 0$. With each activity j its earliest and latest finish time is associated, respectively EF_j and LF_j, calculated as by Kelley and Walker in [2]. The value of LF_{n+1} is set equal to the scheduling horizon H. The horizon H never exceeds the sum of all activity durations.

The following formulas (4.1)–(4.5) define the problem:

$$\min \sum_{t=EF_{n+1}}^{LF_{n+1}} t x_{n+1,t} \tag{4.1}$$

$$\sum_{t=EF_j}^{LF_j} x_{jt} = 1, \quad \text{for } j = 0, \ldots, n+1 \tag{4.2}$$

$$\sum_{t=EF_i}^{LF_i} t x_{it} \leq \sum_{t=EF_j}^{LF_j} t x_{jt} - d_j, \quad \text{for all } (i,j) \in P \tag{4.3}$$

$$\sum_{j=1}^{n} \sum_{q=max\{t,EF_j\}}^{min\{t+d_j-1,LF_j\}} r_{jk} x_{jq} \leq R_k, \quad \text{for } k = 1, \ldots R, \ t = 1, \ldots H \tag{4.4}$$

$$x_{jt} \in \{0,1\}, \quad \text{for } i = 0, \ldots, n+1, \ t = EF_j, \ldots, LF_i \tag{4.5}$$

Constraints (4.2) ensure that each activity is completed exactly once. Precedence constraints are represented by inequalities (4.3). P denotes the set of all pairs of activities (i,j) such that i directly precedes j. Constraints (4.4) guarantee that no more than the available number of units of each resource is required in any time period. Constraints (4.5) means that we consider binary decision variables. The solution of the problem (4.1)–(4.5) is an optimal schedule represent by a list of activity completion times.

The RCPSP, as a generalization of the classical job shop scheduling problem, belongs to the class of strongly NP-hard optimization problems (see Błażewicz et al. [3]). The considered problem class is denoted as $PS|prec|C_{max}$ according to the classification of Brucker et al. [4] or it is denoted as $m, 1|cpm|C_{max}$ according to the classification scheme of Demeulemeester and Herroelen [5].

The project instance of RCPSP is usually represented by an Activity On Node (AON) network. A schedule is defined by the sequence of activity start (completion) times. Sequences of activities are build using different priority rules, e.g.:

- shortest duration first/Shortest Processing Time first (SPT),
- longest duration first/Longest Processing Time first (LPT),
- Earliest Start Time first (EST),
- Earliest Finish Time first (EFT),
- Latest Start Time last (LST),
- Latest Finish Time last (LFT),
- Earlier Due Dates (EDD).

To generate a schedule from the sequence, the so-called Schedule Generation Scheme (SGS) is used. There are three well known SGS: serial, parallel and combined. There are three methods of placing activities in the schedule: forward, backward and from both directions called forward-backward. To improve a solution after applying the backward scheduling, Local Left Shift (LLS) procedure is usually used.

4.2 State of the Art Review

The RCPSP has been analyzed from over 40 years and attracted a lot of attention due to its practical applications. The RCPSP models form actually a big family, most of which belong to the NP-hard class of combinatorial optimization problems.

In this section a short review of the state of the art with respect to the classical RCPSP formulated in Sect. 4.1 is presented. Methods for solving RCPSP are usually divided into three groups: exact, heuristic and metaheuristics.

Surveys of different approaches, computational results obtained by the best methods and their comparisons may be found in several papers. Icmeli et al. in [6] present a survey of project scheduling problems since 1973. It includes the work done on RCPSP, Time/Cost Trade Off (TCTO) problem and Payment Scheduling Problem as well as their combinations. Elmaghraby in [7] reviews the subject of project representation as activity networks. Özdamar et al. in [8] present the research on the RCPSP classified according to the specified objectives and constraints. A comprehensive survey of the RCPSP and its extensions using branch-and-bound procedures is presented by Herroelen et al. in [9]. Brucker et al. in [4] provide a classification scheme, propose a unifying notation and the review of exact and heuristic algorithms for RCPSP, MRCPSP, TCTO, and RCPSP/max problems with other objectives than makespan minimization and, for problems with stochastic activity durations. One of the most interesting reviews was offered by Hartmann and Kölisch in [10] and later updated in [11]. Authors consider heuristic algorithms for the RCPSP, summarize the basic components of heuristic approaches and discuss features of good heuristics based on the computational experiment results. They analyze the behavior of the heuristics with respect to their components such as priority rules and metaheuristic strategy and examine the impact of problem characteristics such as project size and resource scarceness on the performance. In [12] Kölisch and Padman reviewed a vast literature in all areas of project scheduling and management with a perspective that integrates models, data, and optimal and heuristic algorithms for the major classes of project scheduling problems. The state of the art reviews with respect to solving numerous project scheduling problems and the following classification scheme was proposed by Demeulemeester and Herroelen in [5]. A recent literature review on project scheduling problems and variants was provided by Hartmann and Briskorn [13]. Agarwal et al. [14] reviewed various metaheuristic approaches to the RCPSP. Reviews of the RCPSP and its extensions, as well as solution methods, can be also found in [14–19].

Exact methods applied to the RCPSP can be classified into mathematical programming i.e. zero-one programming and implicit enumeration including dynamic programming and branch-and-bound (B&B). Examples of research involving the use of mathematical programming can be found in the papers of Bowman [20], Brand et al. [21], Pritsker et al. [1], Patterson and Roth [22], Decro et al. [23], and Icmeli and Rom [24]. Solutions based on dynamic programming were proposed, for example by Carruthers and Battersby [25] and Petroviç [26]. Branch-and-bound algorithms for the RCPSP were developed, among others, by: Johnson [27], Schrage [28], Balas [29], Davis and Heidorn [30], Stinson et al. [31], Talbot and Patterson [32], Radermacher [33], Christofides et al. [34], Bartusch et al. [35], Bell and Park [36], Demeulemeester and Herroelen [37, 38], Carlier and Néron [39], Brucker et al. [40], Mingozzi et al. [41], or Dorndorf et al. [42].

Larger size and more complex cases of the RCPSP which, in fact, are encountered in the real life, can not, as a rule, be solved in a reasonable time using exact methods since the RCPSP is NP-hard [3, 13]. Practical solutions for such cases require specialized heuristics or metaheuristics.

Studies on heuristics for solving the RCPSP date back to 1963 when Kelley [43] introduced a Schedule Generation Schema (SGS). Schedule generation schemes (serial, parallel or combined) together with priority rules are the core of most heuristic solution procedures for the RCPSP. Priority rules themselves are used in order to select an activity from the activity set. Priority-rule based heuristics combine one or more priority rules and SGS in order to construct one or more schedules [10]. If only one schedule is generated, the approach is called a single pass method, and if more then one schedule is generated, it is called a multi-pass or X-pass method [10]. There are a lot of research on priority single-pass methods for the RCPSP, for example: Davis and Patterson [44], Cooper [45, 46], Alvarez-Valdes and Tamarit [15], Boctor [47], and Özdamar and Ulusoy [8, 48]. Multi-pass methods include multi-priority rule methods and sampling methods. Multi-priority rule methods combine the SGS with a different priority rule at each iteration [47, 49, 50]. Sampling methods use one SGS and one priority rule to obtain the first schedule. Different schedules are obtained from randomized activity selection [15, 16, 45].

Many metaheuristic methods have been applied to solve the RCPSP. The most commonly used are: genetic algorithms (GA), simulated annealing (SA), tabu search (TS), particle swarm optimization (PSO) and ant colony optimization (ACO), as well as non-standard methods combining different approaches. Several efficient approaches are based on using genetic algorithms, as for example: Leon and Rammamoorthy [51], Alcaraz et al. [52, 53], Hartmann [54, 55], Coelho and Tavares [56] or Debels and Vanhoucke [57]. Hybrid GA and other population-based approaches were proposed by Valls et al. in [58, 59]. Agarwal et al. [14, 60] proposed hybrid of genetic algorithm (GA) and neural-network (NN) algorithm as NeuroGentic approach. TS algorithms developed, among others, Thomas and Salhi [50], Nonobe and Ibaraki [61] and Artigues et al. [62]. Simulated annealing (SA) based approaches were proposed by Boctor [63], Cho and Kim [64], Bouleimen and Lecocq [65] or Valls et al. [66]. Approaches based on particle swarm optimization (PSO) were developed by Zhang et al. [67] or Chen et al. [68]. The first applications of ACO to the RCPSP is credited to Merkle et al. [69] and Herbots et al. [70]. Hybrid metaheuristic that combines ACO, GA and local search strategy was proposed by Tseng and Chen [71]. Artificial bee colony (ABC) based approaches were suggested by Akbari et al. [72], Ziaratia et al. [73] or Jia and Seo [74]. Evolutionary algorithm which combines GA, path relinking (PR) and TS was due to Kochetov and Stolyar [75]. Debels et al. [76] combined elements from scatter search (SS), a population-based evolutionary search method and electromagnetism metaheuristic.

Due to the fact that the RCPSP is one of the most intractable and widely researched problem among classic operations research problems, it has always attracted new algorithms and innovative techniques. Among them, during recent years, notable successes can be attributed to swarm optimization, population based and hybrid methods, as well as agent based approaches. Paraskevopoulos et al. in [77] proposed an event list-based evolutionary algorithm. Wang and Fang [78] proposed hybrid estimation of distribution algorithm for the RCPSP. In [79] the same authors described heuristic based on Shuffled Frog-Leaping Algorithm (SFLA) and they presented a speedup method for evaluating new solutions based on some theoretical analysis. Yannibelli and Amandi in [80] proposed a hybrid method by integrating the SA into EA. Evolutionary programming based approaches were proposed by Sebt et al. [81]. Zamani in [82] proposed a competitive GA with a magnet-based crossover operator. Fahmy et al. in [83] proposed PSO with stacking justification for solving the RCPSP.

In the above discussed approaches eventual progress in the use of optimal, heuristic and meta-heuristic methods for the RCPSP has been validated in most cases using two standard problem sets: the Patterson [84] data set and the data set generated by Kölisch et al. [85]. Both benchmark data sets are available in PSPLIB. On the other hand many methods are tested on different benchmark datasets using different stopping criteria. Hence, a comparison between these procedures is, in many cases, difficult.

4.3 Agent-Based Approaches to Solving RCPSP

A few MAS implementations have been proposed for the single-mode Resource-Constrained Project Scheduling Problem (RCPSP). The short survey of some approaches can be found in [86] including methods for solving the classical RCPSP, dynamic, distributed, multi-mode and multi-project problems. Additionally, the approaches for the MRCPSP, described in Sect. 5.3, may be used for the RCPSP as well.

Agent-based approaches to the classical RCPSP are described in the following part of this section.

Shu-Guang et al. [87] proposed the multiple-agent system based on general equilibrium market mechanism to solve the RCPSP and verified it experimentally. In their approach the MAS is combined with distributed decision making method to handle the distributed and dynamic nature of the project scheduling problems. It is supposed to achieve market equilibrium and resource allocation via bidding between activity agents and resource agents.

Wauters et al. [88, 89] proposed a network of distributed reinforcement learning agents for the MRCPSP and tested it for the RCPSP using test set j120 from PSPLIB [90, 91] (see Sect. 4.4.1), containing 480 instances of 120 activities. The approach is described in Sect. 5.3. The computational experiments show that the results are comparable with the best results known from the literature and the approach proposed for the MRCPSP is also a useful tool for solving instances of the RCPSP.

Horenburg et al. [92] presented the MAS for the classical RCPSP with agents for each resource and process. It is an enhancement of the Knotts et al. [17] agent-based scheduling method for the MRCPSP described in Sect. 5.3. In the approach both process agents and resource agents are autonomous and participate in the decision making. Each process agent represents an activity and uses information about the activity and information from the system to manipulate its state. Five stages for the process agents are considered: blocked, admitted, accepted, active, and complete. Resource agents are able to measure their qualification for solving the corresponding task by estimating values of the utility functions and variables. Three stages for solving them are considered: free, active and reserved. Agents which satisfy the specific requirements can negotiate using the central blackboard. Process agents request proposals on the blackboard and resource agents offer their renewable or non-renewable resources. Specific communication protocols have been implemented in all agents as well as in the blackboard. Additionally, to improve the quality of computed results the capital of individual process agent is calculated. This capital decides on the prioritization during bidding procedures.

The multi-agent system proposed by Horenburg et al. [92] was modeled in discrete-event-simulation (DES) and evaluated experimentally using the RCPSP instances from PSPLIB including projects with 30 and 120 activities (j30 and j120). The results are compared with the constrained-based method using Monte-Carlo-simulation (MCS). The agent-based approach has proven to be the better one.

Zheng and Wang in [93] proposed the Multi-Agent Optimization Algorithm (MAOA) for solving the RCPSP. The proposed algorithm uses multiple agents working in a grouped organization environment. Each agent represents a feasible solution. Each group consists of the same number of agents. The best agent in the group is elected as the leader. To share information between groups the second best agent in each group (active agent) is exchanged with the worst agent in the group with the best leader (elite group).

The evolution of agents in MAOA is achieved by using four main elements: social behavior (global and local), autonomous behavior, self-learning, and environment adjustment (migration among groups). Agents are initialized using the regret-based biased random sample method with the LFT priority rule [54] and serial SGS. After initialization they are divided into fixed number of groups. Social behavior i.e. coordination and cooperation between agents include global social behavior where the leader of the elite group cooperates with all the leaders in other groups, and local social behavior where the leader of each group interacts with other agents in the same group. As a social global behavior the magnet-based crossover (MBCO) [82] was used. As local social behavior the Resource-Based Crossover (RBCO) proposed by Fang and Wang [79] was adopted. To realize the autonomous behavior each agent exploits its neighborhood using permutation based swap (PBS) as a local search, and then accepts the new neighbor with a better quality. Agents can be also improved by learning from the acquired knowledge (self-learning). In MAOA for the best leader the SGS forward-backward improvement was used. To share the information among agents the environment is adjusted every 10 generations.

Authors investigated the key parameters of MAOA and performed computational experiment using PSPLIB test sets of j30, j60 and j120. Comparisons of the results with the 14 existing algorithms demonstrates the effectiveness of MAOA in solving the RCPSP.

4.4 A-Teams Solving the RCPSP

In this section six variants of A-Teams implemented for solving the RCPSP are presented. All A-Teams have been constructed using JABAT environment. For each A-Team variant the computational experiment has been carried out and its results have been presented. In Sect. 4.4.1 the first two approaches with a single A-Teams using the static cooperation strategy have been described. Details of the computational experiments also described in this section are identical in all the discussed cases. In Sect. 4.4.2 the essential modifications of the first approaches are indicated and the algorithms used in the further A-Teams implementations are presented. Team of A-Teams based approach where multiple A-Teams were used is presented in Sect. 4.4.3. The next three subsections include descriptions of single A-Teams with dynamic cooperation strategies based on: reinforcement learning (RL) in Sect. 4.4.4, Population Learning in Sect. 4.4.5, and integration of the best rules from previous approaches in Sect. 4.4.6. In each case the computational experiments results follow. Finally, in Sect. 4.4.7 the summary of the proposed approaches are presented including comparison of results with the best one known from the literature.

4.4.1 Single A-Teams with the Static Cooperation Strategies

The first attempt of using A-Team concept for solving the RCPSP dates back to 2007. The approach proposed in [94] includes implementation of classes representing the single mode RCPSP instances: smData and smTask, smSolution, smActivity, and smResource. The classes smData, smTask and smSolution are inherited from the general classes available in JABAT Data, Task, and Solution, respectively. The smData class identifies one test set from PSPLIB storing the text and values. In this class text processing functions are implemented that generate JABAT data structures. The smTask identifies an instance, which attributes include a list of activities $[0, \ldots, n+1]$, and a list of available renewable resources. The smSolution class describes a solution and includes the ordered list of activities (AL) and starting times of the activities obtained from SGS, as well as processing functions, for example moving activities, generating SGS or allocating resources. The remaining classes represent the problem structures. The smResource class identifies renewable resources, storing the value representing a number of the resource units. The smActivity identifies activity, which attributes include the activity number, a list of available renewable resources required by the activity, and a list of predecessors and successors.

Five optimization algorithms solving single mode RCPSP were implemented and used as optimization agents. There are as follows:

- Local Search Algorithm (smLSA),
- Tabu Search Algorithm (smTSA),
- Crossover Algorithm (smCA),
- Minimal Critical Sets based Algorithm (smMCSA),
- Precedence Tree Algorithm (smPTA).

smLSA is a simple local search algorithm which finds the local optimum by moving each activity to all possible places in the solution. Only feasible solutions are acceptable. The best solution found is remembered.

smTSA is a tabu search algorithm where the neighborhood of the initial solution is searched by performing moves that are not tabu. In considered TSA the move rely on two activities exchange. The selected moves are remembered on tabu list. The best solution found is remembered.

smCA is based on the one-point crossover operator. The crossover operation is applied on each pair of solutions from the population until the better solution will be found or all crossing points will be tried unsuccessfully.

smMCSA is based on the approach proposed in [95]. The possible constraints are detected based on Minimal Critical Sets (MCS) and adapting shaving method. Next, the solution respecting constraints is created. The procedure is repeated for the fixed number of iterations.

smPTA is based on the precedence tree approach proposed in [96]. It finds an optimum solution by enumeration for a partition of the schedule consisting of some activities. Next, it finds the solutions of the successive partitions shifted for a fixed step. The best solution found is remembered.

The optimization algorithms described above are used in the proposed A-Team implementation. The following optimization agents derived from OptiAgent class are used:

- OAsmLSA - implementing the Local Search Algorithm with simple shifting move smLSA,
- OAsmTSA - implementing the Tabu Search Algorithm with exchange move smTSA,
- OAsmCA - implementing the Crossover Algorithm smCA,
- OAsmMCSA - implementing the Minimal Critical Sets Algorithm smMCSA,
- OAsmPTA - implementing the Precedence Tree Algorithm smPTA.

All optimization agents (OptiAgents) work in parallel improving solutions from their A-Team common memory managed by the SolutionManager. An individual is represented as a schedule of activities S. The final solution is obtained from the schedule by forward Schedule Generation Scheme (SGS) procedure (see Sect. 4.1).

In the above implementation the basic cooperation strategy for the RCPSP was implemented and used according to the following rules:

- The initial population is generated randomly.
- The individuals from the population are chosen randomly and immediately send to optimization agents,
- An improved solution replaces the worst one from the population,
- The computation for one problem instance in this approach is interrupted after 5 mins.

Another approach based on a similar basic cooperation strategy proposed in [97] slightly differs with respect to membership in the set of optimization agents and cooperation strategy elements. In the set of optimization agents instead of the least effective OAsmMCSA the OAsmPRA is used.

smPRA is a path-relinking algorithm (PRA). For a pair of solutions from the population a path between them is constructed. Next, the best of the feasible solutions from the path is selected. To construct the path of solutions the activities are moved to other possible places in the schedule, as in the case of smLSA. The path-relinking algorithm smPRA is implemented in the optimization agent OAsmPRA.

Because the discussed algorithms have been undergoing several changes over various stages of development, final detailed versions used in the following computational experiment are described further on in Sect. 4.4.2.

A change in the cooperation strategy is that an improved solution replaces the original one, instead of the worse one, from the population (common memory). In the computational experiments presented in [97] the computation for a single problem instance was interrupted after 50 solutions which have proven to be not better then the best one stored in the common memory.

The first experiments have been performed for single A-Team with the static cooperation strategy (SC Strategy) described above. In the following part of this work the SC Strategy using smLSA, smCA, smPTA, smTSA and smMCSA is called SC1

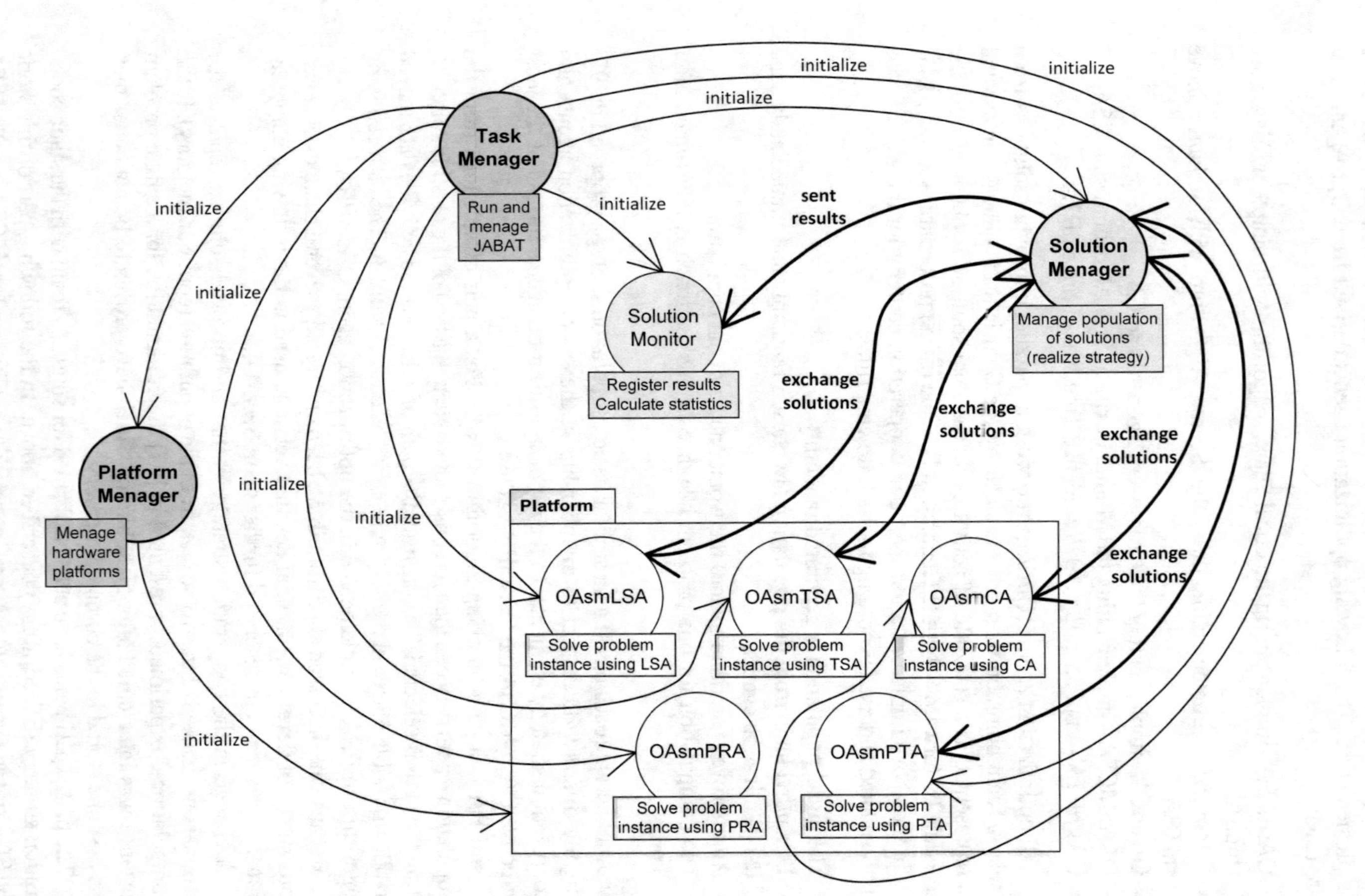

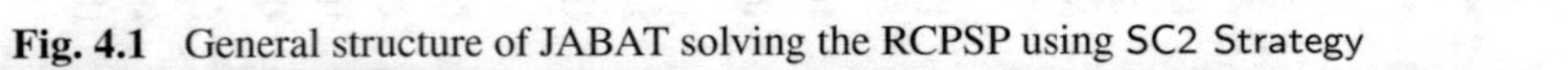
Fig. 4.1 General structure of JABAT solving the RCPSP using SC2 Strategy

Strategy and the SC Strategy using smLSA, smCA, smPTA, smTSA and smPRA is called SC2 Strategy. The general structure of JABAT solving the RCPSP using SC2 Strategy is shown in Fig. 4.1.

To evaluate the effectiveness of the above described A-Teams in solving the RCPSP instances the computational experiments have been carried out. To assure comparability of A-Teams performance all approaches to solving the RCPSP presented in Sect. 4.4 have been validated experimentally using benchmark instances test sets of the RCPSP from PSPLIB. The following sm (single mode) test sets are used: j30 (30 activities), j60, j90, and j120. Each of the first three sets includes 480 problem instances, while set j120 includes 600. The experiments involved computation with a fixed number of optimization agents, fixed population sizes, and fixed stopping criteria.

During all experiments the following characteristics of the computational results have been calculated and recorded: Mean Relative Error (Mean RE) and Maximal Relative Error (Max RE) calculated as the deviation from the optimal (opt) solution for j30 test set or from the best known (bk) solution for j60, j90, and j120 sets and the Critical Path Lower Bound (CPLB), Mean Computation Time (Mean CT) which has been needed to find the best solution and Mean Total Computation Time (Mean TCT) which has been needed to stop all optimization agents and the whole system. Minimal relative error has been found to be 0 in all cases. Each instance has been solved five times and the results have been averaged over these solutions.

All experiments have been carried out using nodes of the cluster Holk of the Tricity Academic Computer Network built of 256 Intel Itanium 2 Dual Core 1.4 GHz with 12 MB L3 cache processors and with Mellanox InfiniBand interconnections with 10 Gb/s bandwidth.

The results presented in this book are recalculated in respect of the number of generated schedules (SGS procedure calls, see Sect. 4.1), which have been limited to 5000. Some of the algorithms implementations have been modified and improved in respect of the language structures.

The results obtained for the two first approaches to solve the RCPSP by the A-Team implemented in JABAT are summarized in Tables 4.1 and 4.2. It can be seen that the first approach is most effective for larger problems with 90 and 120 activities but in the second one the computational time has been reduced significantly.

Table 4.1 Experiment results for the A-Team with SC1 Strategy solving the RCPSP

Number of activities	Mean RE from opt*/bk solution [%]	Max RE from opt*/bk solution [%]	Mean RE from CPLB [%]	Max RE from CPLB [%]	Mean CT [s]	Mean TCT [s]
30	0.46*	9.09*	13.99	120.83	88	127
60	0.76	8.86	11.54	112.99	98	142
90	1.58	16.85	11.25	98.31	88	119
120	2.89	28.09	33.21	201.01	105	150

Table 4.2 Experiment results for the A-Team with SC2 Strategy solving the RCPSP

Number of activities	Mean RE from opt*/bk solution [%]	Max RE from opt*/bk solution [%]	Mean RE from CPLB [%]	Max RE from CPLB [%]	Mean CT [s]	Mean TCT [s]
30	0.44*	22.22*	13.89	127.08	11	34
60	0.70	20.37	11.28	106.49	24	62
90	1.94	10.34	12.03	111.86	40	64
120	3.11	12.5	34.10	207.07	91	122

4.4.2 Algorithms Used in the Further A-Team Approaches

During the first experiments some problems and disadvantages of the environment, strategies and optimization agents have been observed which results in a few modifications. In this section the algorithms used in all subsequent JABAT based A-Team approaches are presented.

Observing the performance of the A-Team with SC Strategy it was easy to note that optimization agents used occurred quite a burdening for the computation. Both, the complexity of algorithms which results in the extended computation time and the number of messages exchanged between agents and common memory significantly influenced the obtained results. Thus, the number of agents as well as their algorithms were modified.

The modified implementations of algorithms for the RCPSP arose from the A-Team approaches proposed for the RCPSP/max developed from 2008 to 2011. These algorithms are described in [98] for the RCPSP/max and in [99–101] for the RCPSP.

To describe the algorithms and present their current versions in pseudocodes the following auxiliary functions are used:

- minCT - requires two schedules (individuals) as arguments and returns the better one with respect to the project complexion time (makespan) (Fig. 4.2);
- smMakeMove - moves the activity form one position in the schedule (activity list) to the another position and returns the modified schedule as a result (Fig. 4.3);
- smReverseMove - reverses the move of the activity form one position to another in the schedule and returns the previous schedule as a result (Fig. 4.4);

```
minCT(S1,S2)
{
    S = S1
    if(CT(S2) < CT(S1))  S = S2
    return S;
}
```

Fig. 4.2 Pseudocode of the minCT function

```
smMakeMove(S, pi, pj)
{
    S0 = S
    Move activity from position pi to position pj in S
    if(precedence and resource constraints in S are satisfied)
        return S
    else return S0
}
```

Fig. 4.3 Pseudocode of the smMakeMove function

```
smReverseMove(S, pi, pj)
{
    Move activity from position pj to position pi in S
    return S
}
```

Fig. 4.4 Pseudocode of the smReverseMove function

```
smMakeExchange(S, pi, pj)
{
    S0 = S
    Exchange activities from positions pi and pj in S
    if(precedence and resource constraints in S are satisfied)
        return S
    return S0
}
```

Fig. 4.5 Pseudocode of the smMakeExchange function

```
smReverseExchange(S, pi, pj)
{
    Exchange activities from positions pi and pj in schedule S
    return S
}
```

Fig. 4.6 Pseudocode of the smReverseExchange function

- smMakeExchange - exchanges the activities form two given positions in the schedule and returns the modified schedule as a result (Fig. 4.5);
- smReverseExchange - reverses the exchange of two activities from given positions in the schedule and returns the previous schedule as a result (Fig. 4.6).

The pseudocodes of the optimization algorithms are presented in Figs. 4.7, 4.8, 4.9 and 4.10.

The smLSA is a local search algorithm which finds local optimum by moving each activity to all possible places in the schedule (in smLSAm) or by exchanging pairs of activities (in smLSAe). For each combination of activities the value of possible solution is calculated. The best schedule is remembered and finally returned. The pseudocode of the smLSAm is shown in Fig. 4.7. The smLSAm algorithm has

```
smLSAm (S, itNumber, iStep, fStep)
{
    bestS = S
    it = 0
    while(it < itNumber)
    {
        for(step = iStep; step < fStep; step = step + 1)
        {
            pi = ++pi % (n - step) + 1
            for (pj = pi; pj < n - step; pj = pj + step)
            {
                smMakeMove(S, pi, pj)
                SGS(S)
                if(CT(S) < CT(bestS)) bestS = S
                smReverseMove(S, pi, pj)
            }
        }
        return bestS
    }
}
```

Fig. 4.7 Pseudocode of the smLSAm algorithm

four parameters. The first, *S* denotes an initial schedule (individual). The second, *itNumber* is the number of iterations for the algorithm. The last two, *iStep* and *fStep* indicate the initial and final distance between activities under moving or exchange. The pseudocode of the smLSAe algorithm differs mainly in using the smMakeExchange instead of smMakeMove and the smReverseExchange instead of smReverseMove function, hence it is not presented separately.

The smTSAe algorithm is an implementation of tabu search metaheuristic [102–104]. It finds local optimum by exchanging each two activities in the schedule, for which such exchange is possible. The best exchange move from the neighborhood of the solution, which is not tabu, is chosen and performed. The pseudocode of the smTSAe is shown in the Fig. 4.8. The algorithm smTSAe requires four parameters. The first, *S* denotes an initial schedule (individual). The second, *pi* indicates the activity position from which the exchange starts. The parameter *maxItNWI* denotes a maximum number of iterations in which no improvement is found. The last, *step* denotes the distance between activities under exchange. To exchange activities the smMakeExchange and smReverseExchange functions are used in the smTSAe. In smTSAm the simple shifting move, instead of exchange move, for the neighborhood searching is used. Hence, the smMakeMove and smReverseMove function are used, respectively.

The tabu list *TabuList* has been implemented to store the information of recent moves which should not be repeated by a fixed number of iterations. For example:

- Setting the exchange move (pi, pj, n) as the tabu one prevents from repeating the same exchange by n iterations. It blocks all moves that exchange activity from position *pi* with activity from position *pj* for n iterations.

```
smTSAe(S, pi, maxItNWI, step)
{
    TabuList = ∅
    it = 0
    bestS = S
    while(it < maxItNWI)
    {
        bestSit = null
        for(pj = pi+1; pj < n-1; pj = pj+step)
        {
            move = (S, pi, pj)
            if(move is not in TabuList)
            {
                smMakeExchange(move)
                SGS(S)
                if(CT(S) < CT(bestSit)) bestSit = S
                smReverseExchange(move)
            }
        }
        update TabuList
        if(bestSit is not null)
        {
            if(CT(bestSit) < CT(bestS))
            {
                bestS = bestSit
                it = 0
            }
            Add moves to TabuList:
                (pi, pj, maxItNWI)
                (null, pi, ⌊n/4⌋)
                (pi, null, ⌊n/4⌋)
                (pj, null, ⌊maxItNWI/2⌋)
        }
        else it--
        pi = pi % (n-2) + 1
    }
    return bestS
}
```

Fig. 4.8 Pseudocode of the smTSAe algorithm

- Setting the exchange move (pi, $null$, $\lfloor n/4 \rfloor$) as the tabu one blocks exchanges of activity from position pi with any other activity for $\lfloor n/4 \rfloor$ iterations.

The best schedule is remembered and finally returned.

The smCA is an algorithm based on the idea of the one point crossover operator. For a pair of solutions $S1$ and $S2$ one point crossover is applied. The parameter *step* determines the frequency the operation is performed. The best schedule is remembered and finally returned. The pseudocode of the smCA is shown in Fig. 4.9.

The smPRA is an implementation of the path-relinking algorithm [105, 106]. For a pair of solutions a path between them is constructed. The solutions $S1$ and $S2$

```
smCA(S1, S2, step)
{
    bestS = minCT(S1,S2);
    for (pi = 1;  pi ≤ n;  pi = pi + step)
    {
        Do simple (one point) crossover on S1 and S2 in point pi;
        bestS = minCT(bestS, minCT(S1,S2));
    }
    return bestS;
}
```

Fig. 4.9 Pseudocode of the smCA algorithm

are required as the algorithm parameters. The path consists of schedules obtained by carrying out a single shifting move from the preceding schedule. The move is performed by smMakeMove function (see Fig. 4.3). For each schedule in the path the value of the respective solution is checked. The best schedule is remembered and finally returned. The pseudocode of the smPRA is shown in Fig. 4.10.

Four optimization algorithms originally proposed in [101] are used in all subsequent A-Teams implementations described in this section. There are: smLSAm, smTSAe, smCA, and smPRA (see Figs. 4.7, 4.8, 4.9 and 4.10). The following optimization agents derived from OptiAgent class are used:

- OAsmLSAm - implementing the Local Search Algorithm with simple shifting move smLSAm,
- OAsmTSAe - implementing the Tabu Search Algorithm with exchange move smTSAe,
- OAsmCA - implementing the Crossover Algorithm smCA,
- OAsmPRA - implementing the Path Relinking Algorithm smPRA.

Similarly to the optimization algorithms and optimization agents the cooperation strategy has been evolving during the research development. The modified cooper-

```
smPRA(S1, S2)
{
    bestS = minCT(S1,S2);
    S = S1
    for(pi = 1;  pi ≤ n;  pi++)
    {
        Denote by ai the activity from position pi in the schedule S2
        pj = position of the activity ai in the schedule S
        smMakeMove(S, pj, pi)
        if(CT(S) < CT(bestS))  bestS = S
    }
    return bestS
}
```

Fig. 4.10 Pseudocode of the smPRA algorithm

ation strategy foundations arose from the A-Team approaches proposed for RCPSP and RCPSP/max developed from 2007 to 2011. The details of the basic static cooperation strategy with blocking can be found in Sect. 4.4.4 and [101].

4.4.3 Randomized Team of A-Teams with Static Cooperation Strategy

The idea of multiple A-Teams approach originated in [99]. In that paper the JABAT Team of A-Teams (TA-Teams) approach for solving the RCPSP was proposed and experimentally evaluated. Several modifications and improvements with respect to optimization agents and strategies have been implemented within a dedicated TA-Teams architecture and used to solve instances of the RCPSP problem. In the discussed approach more then one A-Team was employed resulting in several SolutionManagers and several populations acting in parallel. Hence, computations involve a number of optimization agents. As it was mention before increased number of solutions and optimization agents increases the number of messages exchanged between agents and common memory. In the agent-based environments the significant part of computation time is used for agent communication which has an influence on both - overall computation time and quality of the results.

The optimization/improvement algorithms are described in the previous section and used to solve the RCPSP instances by Team of A-Teams. In this section the implementation with eight A-Teams and randomized interaction strategy is described.

The cooperation strategy within one A-Team managed by one SolutionManager is based on the following rules:

- All individuals in the initial population of solutions are generated randomly and stored in the common memory.
- Individuals which are to be improved are selected from the common memory randomly and blocked, which means that once selected individual (or individuals) cannot be selected again until all other individuals have been tried.
- Returning individual replaces the blocked one or the worse from the blocked ones. If the returning individual is not better than the best one from the population by a fixed number of iterations (iteration is understood as receiving improved individual from optimization agent) then the worst individual in the common memory is replaced by a randomly generated one, representing a feasible solution.
- The computation time of a single A-Team is controlled by the no improvement time gap set by the user. If in such time gap no improvement of the current best solution has been achieved, the A-Team stops computations.

In case of the TA-Teams approach the interaction strategy describing interactions between A-Teams is defined additionally. Interaction strategy controls the migration of solutions between A-Teams. In this approach it is based on the so called randomized strategy. Each SolutionManager may ask for a new solution and randomly

Table 4.3 Experiment results for team of A-Teams with RMSC Strategy solving the RCPSP

Number of activities	Mean RE from opt*/bk solution [%]	Max RE from opt*/bk solution [%]	Mean RE from CPLB [%]	Max RE from CPLB [%]	Mean CT [s]	Mean TCT [s]
30	0.02*	1.75*	13.40	120.83	14	67
60	0.54	8.45	11.12	103.90	32	70
90	0.95	9.68	10.81	110.17	38	66
120	2.57	20.00	33.18	204.04	73	128

selected other SolutionManager sends it to the first one. The SolutionManager asks for a new solution when the current best solution in its population (common memory) has not been changed for a fixed part of no improvement time gap. Further details of the Team of A-Teams approach can be found in [99].

In Table 4.3 the results for the Team of eight A-Teams, with static cooperation strategy and randomized interaction strategy, working in parallel are presented. It can be noted that the Team of A-Teams approach produces significantly better results in term of mean relative errors than approaches based on a single A-Team. The maximal relative errors are significantly lower for the test sets j30 and j60, but higher for the test set j120. The computation times are slightly worse. Moreover, development, implementation and testing of such a system is much more complicated and time consuming as well.

4.4.4 *A-Team with the Dynamic Cooperation Strategy with Reinforcement Learning*

Further improvements of the A-Team performance with respect to solving the RCPSP can be achieved through modification and improvements of the cooperation strategy. In [101, 107, 108] studies aiming at finding more effective strategies for A-Team have been undertaken. It appears that a significant performance improvement is possible with introduction of the dynamic strategy approach. Strategies described in Sects. 4.4.1 and 4.4.3 are static, that is constant within a single computation cycle. In this subsection a dynamic interaction strategy is discussed. Together with replacing cooperation strategy some further modification of optimisation/improvement algorithms in case of the smLSA and smTSA may improve A-Team performance.

All further discussed strategies use concept of blocking. The basic static blocking cooperation strategy respects the following rules:

- All individuals in the initial population of solutions are generated randomly and stored in the common memory.
- Individuals for improvement are selected from the common memory randomly and blocked which means that once selected individual (or individuals) cannot be selected again until the OptiAgent to which they have been sent returns the solution.

- The returning individual, which represent a feasible solution, replaces its original version before the attempted improvement. It means that the solutions are blocked for the particular OptiAgent and the returning solution replaces the blocked one or the worst from the blocked once. If none of the solutions is worse, the random one is replaced. All solutions blocked for the considered OptiAgent are released and returned to the common memory.
- A new feasible solution is generated with fixed probability and replaces the random worse one from the population.
- The environment state is calculated and remembered. This state includes the best individual and the population average diversity. The state is calculated every fixed number of iterations. To reduce the computation time, average diversity of the population is calculated by comparison with the best solution only. Diversity of two solutions for the RCPSP problem is calculated as the sum of differences between activities starting times in the projects.
- The A-Team stops computations where the average diversity in the population is less then fixed threshold (e.g. 0.01%).

The idea of using Reinforcement Learning (RL) to make cooperation strategy for the A-Team solving instances of the RCPSP more flexible was first proposed in [101]. RL [109–111] belongs to the category of unsupervised machine learning algorithms. It is described as learning which action to take in a given situation (state) to achieve one or more goal(s).

Reinforcement learning is usually used in solving combinatorial optimization problems at three levels: direct, metaheuristic or hyperheuristic. RL is also used in Multi-Agent Reinforcement Learning (MARL) where multiple reinforcement learning agents act together in the common environment [112, 113]. In [101] RL was used to control the cooperation strategy parameters in the A-Team. In the proposed approach reinforcement learning based on utility values, proposed in [114, 115], is used.

Three learning rules (RL rules) are formulated and integrated with the blocking cooperation strategy. The rules result in updating probability values for various operations performed within an A-Team. As a result, the reinforcement learning is used to control the following elements of the cooperation strategy:

RL1 controls the replacement of one individual from the population by other randomly generated one.
RL2 controls the method of choosing the individual for replacement.
RL3 individuals in the population are grouped according to certain features, and next the procedure of choosing individuals to be forwarded to optimization agents from respective groups is controlled by RL.

Furthermore, the following combinations of the control elements are considered:

RL12 RL1 and RL2,
RL13 RL1 and RL3,
RL23 RL2 and RL3,
RL123 RL1 and RL2 and RL3.

Each single rule, as well as, each of their combinations may be used in a separate dynamic cooperation strategy.

In the RL1 the probability of randomly generating a new solution P_{rg} is calculated directly. The probability is increased in two cases: where the average diversification in population decreases and where it decreases deeply. The P_{rg} is decreased in three environment states: where average diversity in the population increases, where the best solution is found and where the new solution is randomly generated.

In the RL2 different methods of replacing an individual in the population by the new randomly generated individual have probabilities calculated by RL algorithm P_{mr}. The spectrum of considered methods includes the following three:

- new solution replaces the random one in the population,
- new solution replaces the random worse one in the population,
- new solution replaces the worst solution in the population.

The weight w_m for each of the replacement methods is calculated, where $m \in MR$, $MR = \{random, worse, worst\}$. The w_{random} is increased where the population average diversity decreases and it is decreased in the opposite case. The w_{worse} and w_{worst} is decreased where the population average diversity decreases and they are increased in the opposite case. The probability of selection the method m is calculated as

$$P_{mr} = \frac{w_m}{\sum_{i \in MR} w_i}.$$

In the RL3 the similar RL utility values are introduced for the groups of solutions and different kinds of algorithms. The RL is used during solving one instance of the problem. In this case the additional parameters of environment state are remembered as features of each individual (solution) in the population. These are the average number of unused resources in each period of time $\overline{f_{URP}}$ and average lateness of activities $\overline{f_{LA}}$. These two features allow to group solutions from the population and allocate the optimization algorithms to such solutions, for which they are expected to achieve the best results. For each algorithm the matrix of weights allocated to each solution feature is remembered

$$W^{Alg}_{\overline{f_{URP}},\overline{f_{LA}}} = \left[w^{Alg}_{i,j} \right],$$

where $i \in F_{URP}$ and $j \in F_{LA}$. F_{URP} and F_{LA} denote sets of average unused resources in each period of time and average activity lateness, respectively.

The weight is increased where the optimization agent implementing given kind of algorithm returns better or the best solution (positive reinforcement). The weight is decreased where the optimization agent returns not better or worse solution (negative reinforcement). The probability of choosing the optimisation agent is calculated as

$$P^{Alg}_{\overline{f_{URP}},\overline{f_{LA}}} = \frac{W^{Alg}_{\overline{f_{URP}},\overline{f_{LA}}}}{\sum_{i \in F_{URP}, j \in F_{LA}} W^{Alg}_{ij}}.$$

The RL12, RL13, RL23 and RL123 rules are combinations of the RL1, RL2 and RL3 where more then one rule is used. The rules RL1 and RL2 are used consecutively just before adding the new solution to the population after receiving it by the SolutionManager from the OptiAgent. The RL3 rule is used just before reading the solution from the population. The RL3 rule act asynchronously to the RL1 and RL2.

All proposed RL rules as well as their combinations are used to update the probability values for its operations in seven dynamic blocking cooperation strategy implementations. Their performance has been tested experimentally. The results show that the probability values influence the whole strategy performance e.g. population managing and communication with the OptiAgents. In these implementations the effectiveness of changes is checked every fixed number of iteration *itNS* by calculating the environment state. The reward/punishment signal is calculated based on the environment state parameters and used to reward or punish the weights in accordance with the respective learning rule. The details of this approach can be found in [101].

The experiments show that the proposed approaches are effective and promising. The best one is the strategy based on the rule RL123, where all elements are controlled. In this book the strategy is called DCRL Startegy. The results obtained by DCRL Strategy are presented in Table 4.4.

The results obtained for the A-Team with dynamic cooperation strategy based on reinforcement learning DCRL Strategy are comparable with the results obtained for the approach where Team of A-Teams (RMSC Strategy) has been used. The mean relative errors are worse by an average of 0.12%, and more precisely they are equal for the test set j30, and higher by 0.01%, 0.11%, and 0.37% for j60, j90, and j120, respectively. Simultaneously the results are significantly better from the results obtained by the best so far A-Team with SC2 Strategy (Sect. 4.4.1) by an average of 0.41%, in detail by 0.42%, 0.15%, 0.88% and 0.17% for j30, j60, j90 and j120, respectively.

Table 4.4 Experiment results for the A-Team with DCRL Strategy solving the RCPSP

Number of activities	Mean RE from opt*/bk solution [%]	Max RE from opt*/bk solution [%]	Mean RE from CPLB [%]	Max RE from CPLB [%]	Mean CT [s]	Mean TCT [s]
30	0.02	2.06	13.41	122.92	2	25
60	0.55	5.79	11.21	109.09	9	51
90	1.06	7.44	11.01	110.17	21	68
120	2.94	8.97	34.04	207.07	72	185

4.4.5 A-Team with the Dynamic Strategy Based on Population Learning

Population learning algorithm has inspired yet another approach to designing the dynamic cooperation strategy for the A-Team solving instances of the RCPSP.

Population Learning Algorithm (PLA) introduced by Jędrzejowicz in [116] is a population-based metaheuristic inspired by the social education systems in which the process of solving a problem is divided into stages, and a diminishing number of individuals enter more and more advanced learning stages. Hence, at higher stages more advanced and complex learning techniques are used. In each stage the considered optimization problem is solved using a mix of the independent learning procedures. Value of the objective function is directly used as a measure of the quality of individuals and, hence as a selection criterion. After each learning stage the average value of objective function is used to select the individuals.

The idea of using PLA to design dynamic cooperation strategy for A-Team solving the RCPSP was proposed in [107]. It rely on using different optimization agents in different learning stages. The objective function (complexion time) is used to measure the quality of individuals. As the selection criterion an average value of complexion time (makespan) in the population is used avgCT(P). As the stopping criterion the average diversity in the population avgDiv(P) and the number of SGS procedure calls ($nSGS$) are calculated. The general schema of the dynamic cooperation strategy based on the PL paradigm (DCPL Strategy) is shown in Fig. 4.11. The coefficient $\alpha \in [0, 1]$ is used to determine the number of individuals for the next stage.

```
DCPL_Strategy
{
    Generate the initial population P
    Calculate environment state
    for (all learning stages)
    {
        while (none of the stopping criteria are met)
        {
            Use OptiAgents to improve solutions
            Generate a new solution S with probability P_mg
            Replace the individual in P by S with probability P_mr
            Calculate environment state
        }
        Calculate avgCT(P)
        Remove from P all individuals S for which CT(S) < α avgCT(P)
    }
}
```

Fig. 4.11 General schema of the DCPL Strategy

The proposed DCPL Strategy is used to control parameters and to manage the process of searching for solutions of the RCPSP instances by a team of agents. In this approach parameter values depend on the current state of the environment and the current stage of learning. Both undergo dynamic changes during the computation. The basic features of the DCPL Strategy remain similar to the DCRL Strategy described earlier.

To validate the approach three PLA based strategies have been implemented and tested, including 1, 2, or 3 learning stages. In each stage different parameters and optimization agents are used. The main parameters include two probabilities P_{mg} and P_{mr}.

P_{mg} denotes probability of using one of the four available methods for generating a new individual in the population:

- *mgr* randomly;
- *mgrc* applying one point crossover operator for two randomly selected individuals in the population;
- *mgb* randomly changing features of the best individual in the population;
- *mgbc* applying one point crossover operator for two randomly selected individuals from the five best individuals in the population.

The weight w_{mg} for each method of generating is calculated, where $mg \in MG$, $MG = \{mgr, mgrc, mgb, mgbc\}$. The w_{mgr} and w_{mgrc} are increased where the population average diversity decreases and they are decreased in the opposite case. The w_{mgb} and w_{mgbc} are decreased where the population average diversity increases and they are increased in the opposite case. The probability of selecting the method *mg* is calculated as

$$P_{mg} = \frac{w_{mg}}{\sum_{mg \in Mg} w_{mg}}.$$

In case of replacing an individual in the population the same methods as for DCRL Strategy have been used. Thus, the probability P_{mr} and the weights are calculated in the same way (see Sect. 4.4.4).

The sets of settings for PLA based strategy with three learning stages are shown in the Table 4.5. The values in brackets for optimization agents denote the numbers of iterations of the respective optimization algorithms according to their descriptions in Sect. 4.4.2. For the LSA it is the total number of iterations, and for the TSA the maximum number of iterations in which no improvement is found.

The set of optimization algorithms used in this and in the following approach described in this book, includes six optimisation agents: OAsmCA, OAsmPRA, OAsmLSAm, OAsmLSAe, OAsmTSAm and OAsmTSAe.

Computational experiment results for the A-Team with DCPL Strategy and six optimization agents are shown in Table 4.6. The experiment shows that the results are comparable with the results obtained by the A-Team with DCRL Strategy or even outperform them. On average, results are better than the results obtained by DCRL Strategy by 0.23%. More precisely it is equal for instances from the test set j30, better by 0.08% for instances from the test set j60 with 60 activities, 0.17% for

Table 4.5 Parameter values for three learning stages used in DCPL Strategy

Stage 1	
Initial weights	$w_{mgr} = 50$, $w_{mgrc} = 50$, $w_{mgb} = 0$, $w_{mgbc} = 0$ $w_{mrr} = 100$, $w_{mrw} = 0$, $w_{mrt} = 0$
Optimization agents	OAsmCA, OAsmPRA
Selection criteria	$\alpha = 0.5$
Stopping criteria	avgDiv$(P) < 0.1$ or $nSGS > 2000$
Stage 2	
Initial weights	$w_{mgr} = 25$, $w_{mgrc} = 25$, $w_{mgb} = 25$, $w_{mgbc} = 25$ $w_{mrr} = 34$, $w_{mrw} = 33$, $w_{mrt} = 33$
Optimization agents	OAsmLSAm($5n$), OAsmLSAe($5n$)
Selection criteria	$\alpha = 0.5$
Stopping criteria	avgDiv$(P) < 0.5$ or $nSGS > 1500$
Stage 3	
Initial weights	$w_{mgr} = 0$, $w_{mgrc} = 0$, $w_{mgb} = 50$, $w_{mgbc} = 50$ $w_{mrr} = 0$, $w_{mrw} = 20$, $w_{mrt} = 80$
Optimization agents	OAsmTSAm($2n$), OAsmTSAe($2n$)
Selection criteria	$\alpha = 0.8$
Stopping criteria	avgDiv$(P) < 0.05$ or $nSGS > 1500$

Table 4.6 Experiment results for the A-Team with DCPL Strategy for solving RCPSP

Number of activities	Mean RE from opt*/bk solution [%]	Max RE from opt*/bk solution [%]	Mean RE from CPLB [%]	Max RE from CPLB [%]	Mean CT [s]	Mean TCT [s]
30	0.02*	1.47*	13.40	122.92	2	30
60	0.47	5.79	11.08	109.09	13	59
90	0.89	8.06	10.76	110.17	22	68
120	2.27	10.66	32.84	198.59	78	178

instances from the test set j90 with 90 activities and 0.67% for instances from the test set j120 with 120 activities.

4.4.6 *A-Team with Dynamic Cooperation Strategy Based on Integration*

Good performance of both so far discussed dynamic cooperation strategies: DCRL Strategy and DCPL Strategy, stood behind an attempt to integrate them expecting a synergetic effect. The idea has been proposed in [108] where the Dynamic Cooperative Interaction Strategy (DCI Strategy) has been described. To standardize terminology the strategy is called the Dynamic Cooperation Strategy based on the Integration of the best rules from previous approaches, notation remains the same - DCI Strategy.

In this approach the method of generating new individuals is changed. To generate a new individual, also in the initial population, randomly chosen priority rule and serial forward SGS are used. As compared with the earlicst discussed approaches the following main changes in the strategy are made:

- To generate new individuals in the initial population, randomly chosen priority rule and serial forward SGS are used.
- The third probability measure P_{ma} is added. It means probability of selecting the OptiAgent used to improve an individual in the population.

Similarly to the DCPL Strategy, there are P_{mg} and P_{mr} with the same methods. There are four possible methods of generating a new individual in the current population: *mgr*, *mgrc*, *mgb*, *mgbc* (see Sect. 4.4.5).

There are three methods of replacing an individual from the population by a new one considered:

mrr new solution replaces the random one in the population;
mrw new solution replaces the random worse one in the population;
mrt new solution replaces the worst solution in the population.

There are six optimization agents, representing six optimization algorithms, used, as in the previous case: smCA, smPRA, smLSAm, smLSAe, smTSAm and smTSAe. For each of them the weight w_{ma} is calculated, where $ma \in MA$, $MA = \{mac, map, malm, male, matm, mate\}$. The w_{ma} is increased if the optimization agent received the improved solution and is decreased in the other case. Additionally, the weights for *mac* and *map* are increased where the average diversity of the population decreases and they are decreased in the opposite case. The weights for *malm, male, matm, mate* are increased where the average diversity of the population increases and they are decreased in the opposite case.

Probabilities of selecting the respective method are calculated as follows:

$$P_{mg} = \frac{w_{mg}}{\sum_{mg \in M} w_{mg}}, \quad P_{mr} = \frac{w_{mr}}{\sum_{mr \in M} w_{mr}}, \quad P_{ma} = \frac{w_{ma}}{\sum_{ma \in M} w_{ma}}.$$

The current environment state parameters are updated every fixed number of iterations, denoted as *nITns*. Such update includes:

- updating anew w_{mgr}, w_{mgrc}, w_{mgb}, w_{mgbc};
- updating anew w_{mrr}, w_{mrw}, w_{mrt};
- updating anew w_{mac}, w_{map}, w_{malm}, w_{male}, w_{matm}, w_{mate};
- updating the best current solution;
- calculating *nSGS*;
- calculating avgCT(P);
- calculating avgDiv(P).

Table 4.7 Parameter values used in DCI Strategy

Initial weights	$w_{mgr} = 25$, $w_{mgrc} = 25$, $w_{mgb} = 25$, $w_{mgbc} = 25$, $w_{mrr} = 34$, $w_{mrw} = 33$, $w_{mrt} = 33$, $w_{maCA} = 30$, $w_{maPRA} = 30$, w_{maLSAm}=10, w_{maLSAe}=10, w_{maTSAm}=10, w_{maTSAe}=10
Optimization agents	OAsmCA, OAsmPRA, OAsmLSAm($5n$), OAsmSAe($5n$), OAsmTSAm($2n$), OAsmTSAe($2n$)

Table 4.8 Experiment results for the A-Team with DCI Strategy solving the RCPSP

Number of activities	Mean RE from opt*/bk solution [%]	Max RE from opt*/bk solution [%]	Mean RE from CPLB [%]	Max RE from CPLB [%]	Mean CT [s]	Mean TCT [s]
30	0.02*	1.62	13.41	120.83	2	39
60	0.32	6.61	10.86	109.00	15	57
90	0.62	7.14	10.37	103.39	20	59
120	1.89	20.00	32.12	239.60	78	176

The general schema of the proposed DCI Strategy is shown in Fig. 4.12. The *nSGS* denotes the number of SGS procedure calls performed by the algorithm. Further details of the approach can be found in [108].

The effectiveness of the approach has been validated experimentally, and the results are shown in Table 4.8. The DCI Strategy initial settings have been set as follow:

```
DCI_Strategy
{
    Generate the initial population P
    Calculate the environment state
    it=0;
    while(avgDiv(P)>=0.05 and nSGS<5000)
    {
        it++;
        Select OptiAgent with probability P_ma
        Select solutions to improve randomly and remember them in the set S2imp
        Use selected OptiAgent to improve solutions from S2imp
        Generate a new solution S_new with probability P_mg
        Replace the individual selected with probability P_mr in P by S_new
        if(it mod nITns = 0) Calculate the environment state
    }
}
```

Fig. 4.12 General schema of the DCI Strategy

The results show that the A-Team with DCI Strategy proves effective in solving instances of the RCPSP. Integration of the elements from previous approaches results in the best strategy for the RCPSP proposed in this book. The results are in average by 0.18% better than the results for A-Team with DCPL Strategy, and more precisely they are equal for the test set j30 and better by 0.14%, 0.25%, and 0.33% for the test sets j60, j90, and j120, respectively. The reason is that using the most sophisticated algorithms seems beneficial to intensify of exploitation.

4.4.7 Concluding Remarks

Computational experiments results show that proposed A-Team architectures are effective for solving the RCPSP. The Mean Relative Errors (Mean RE) from optimal or best known solutions presented in this section are summarized in Table 4.9. The graphical representation of these results are shown in Fig. 4.13. The most effective approaches are A-Teams with DCPL Strategy and DCI Strategy. In both these approaches six optimization algorithms and six optimization agents representing these algorithms have been used: OAsmCA, OAsmPRA, OAsmLSAm, OAsmLSAe, OAsmTSAm and OAsmTSAe.

To check whether there is a significant difference between the considered A-Teams (i.e. cooperation strategies) the non-parametric Friedman test has been used. It makes possible to answer the question weather particular working strategies are equally effective independently of the kind of problem being solved. Additionally, the mean ranks calculated for each A-Team have been inspected to roughly evaluate their effectiveness.

The test is based on ranks assigned to each A-Team participating in the experiment, i.e. A-Team with SC1, SC2, RMSC, DCRL, DCPL, and DCI Strategy. To assign ranks, the 6 point scale was used, with 6 points for the best and 1 point for the worst result found by the A-Team for a particular problem instance. When the results are identical, the same amount of points equal the mean of consecutive ranks is assigned to each such result (i.e. there are ties among ranks).

The test aimed at deciding among the following hypotheses:

Table 4.9 Mean RE from the optimal (*) or best known solution for JABAT-based A-Team approaches to solve RCPSP discussed in this section

Number of activities	SC1 Strategy [%]	SC2 Strategy [%]	RMSC Strategy [%]	DCRL Strategy [%]	DCPL Strategy [%]	DCI Strategy [%]
30	0.46*	0.44*	0.02*	0.02*	0.02*	0.02*
60	0.76	0.70	0.54	0.55	0.47	0.32
90	1.58	1.94	0.95	1.06	0.89	0.62
120	2.89	3.11	2.57	2.94	2.27	1.89

H_0 zero hypothesis: A-Teams (cooperation strategies) are statistically equally effective i.e. obtain similar results with nonsignificant differences,

H_1 alternative hypothesis: not all A-Teams (cooperation strategies) are equally effective.

The test has been performed on 6 cooperation strategies and 4 data sets of the considered single mode RCPSP from PSPLIB [91]: j30, j60, j90, each including 480 instances, and j120 including 600 instances. The analysis has been carried out at the significance level of 0.05 and 5 degrees of freedom. The respective values of the χ^2 statistics are shown in Table 4.10.

The critical value of χ^2 distribution for the assumed values equals 11.07. Since the obtained values of the statistics χ^2, in all cases, are greater than the critical one, hypothesis H_0 is rejected. Thus, the obtained result proves hypothesis H_1 which claims that not all A-Teams are equally effective in solving the RCPSP instances.

The mean values of ranks, calculated for the considered A-Team, might suggest some ranking with respect to their efficiency. These values are shown in Fig. 4.14.

As it could be seen in Fig. 4.14, four A-Teams have similar ranks and can be considered as superior in case of the instances including 30 and 90 activities. These are three A-Teams with dynamic cooperation strategies (DCRL, DCPL and DCI) and the Randomized Multiple A-Team with Static Cooperation Strategy. In case of

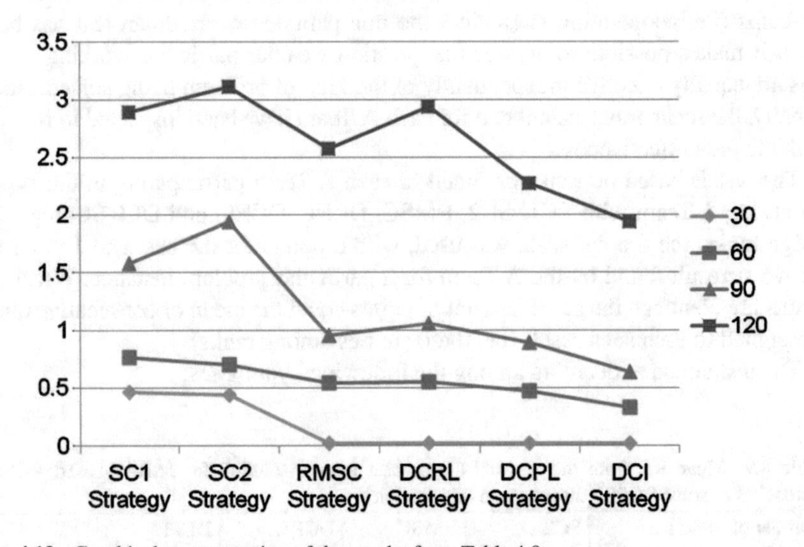

Fig. 4.13 Graphical representation of the results from Table 4.9

Table 4.10 Values of the $\chi 2$ statistics for the RCPSP

Number of activities	χ^2
30	326.25
60	76.64
90	274.49
120	225.83

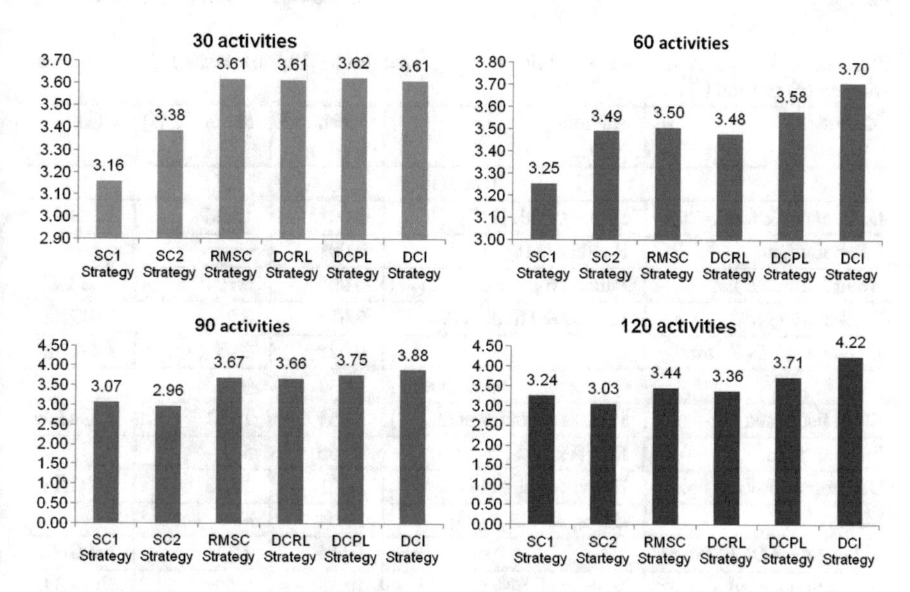

Fig. 4.14 The mean values of Friedman test ranks for the RCPSP

the instances including 60 and 120 activities the A-Team with DCI Strategy can be considered as the superior one.

Results obtained by the proposed A-Teams have been compared with the best results known from the literature. Such comparison, presented in Table 4.11 shows that the proposed approach is effective in solving RCPSP and comparable with the best currently known algorithms. In the presented comparison the other Multi-Agent System MAOA, proposed recently in [93] and described in Sect. 4.3, gives a very similar results and uses some similar assumptions and methods. Agents in MAOA represent feasible solutions and work in organization environment, they also use such methods like crossover and local search. Differences are in grouped organization and strategy in which agents communicate between themselves instead of the common memory.

Others approaches from the literature include: evolutionary algorithms, genetic algorithms and other population-based approaches as well as decomposition, variable neighborhood search and local search. All of presented results use the same test sets from PSPLIB and all of them produce quite similar results. The differences between the Mean RE from optimal results for test set j30 and from CPLB for test sets j60, j90 and j120, reported in the literature and presented in this part of the book are very small: 0.02% for the test set j30, 0.35% for j60 and j90, and 1.86% for j120. However, the detailed and honest comparison is difficult because of the different systems, platforms and machines (processors) used. In respect to the practical applications the differences for the larger projects are more significant, so the proposed approaches for projects with higher number of activities should be tested and compared.

Table 4.11 Comparison with the best literature reported results [11, 60]. Mean RE from optimal solution (*) or from CPLB

Method	Authors	Mean RE [%]	Mean CT [s]	CPU
		30 activities		
Decompos. & local opt	Palpant et al. [117]	0.00*	10.26	2.3 GHz
Filter and fan	Ranjbar [118]	0.00*	–	–
Event list-based EA	Paraskevopoulos et al. [77]	0.00*	0.19	1.33 GHz
VNS-activity list	Fleszar & Hindi [119]	0.01*	5.9	1.0 GHz
A-Team with DCI Strategy		*0.02**	*2.43*	*1.4 GHz*
		60 activities		
Event list-based EA	Paraskevopoulos et al. [77]	10.54	16.31	1.33 GHz
Filter and fan	Ranjbar [118]	10.56	5	–
Decompos. & local opt	Palpant et al. [117]	10.81	38.78	2.3 GHz
MAOA	Zheng & Wang [93]	10.84	–	–
A-Team with DCI Strategy		*10.86*	*15*	*1.4 GHz*
Population–based	Valls et al. [59]	10.89	3.65	400 MHz
		90 activities		
Filter and fan	Ranjbar [118]	10.11	5	–
Population-based	Valls et al. [59]	10.19	9.49	400 MHz
Decompos. & local opt.	Palpant et al. [117]	10.29	61.25	2.3 GHz
Decomposition based GA	Debels & Vanhoucke [60]	10.35	–	–
A-Team with DCI Strategy		*10.37*	*25*	*1.4 GHz*
GA–hybrid, FBI	Valls et al. [120]	10.46	0.61	400 MHz
		120 activities		
Event list-based EA	Paraskevopoulos et al. [77]	30.78	123.45	1.33 GHz
Filter and fan	Ranjbar [118]	31.42	5	–
Population-based	Valls et al. [59]	31.58	59.43	400 MHz
A-Team with DCI Strategy		*32.12*	*86*	*1.4 GHz*
Decompos. & local opt.	Palpant et al. [117]	32.41	207.93	2.3 GHz
MAOA	Zheng & Wang [93]	32.64	–	–

References

1. Pritsker, A.A.B., Watters, L.J., Wolfe, P.M.: Multi-project scheduling with limited resources: a zero-one programming approach. Manag. Sci. **16**(1), 93–108 (1969)
2. Kelley, J.E. Jr., Walker, M.R.: Critical path planning and scheduling. In: Proceedings of the Eastern Joint Computer Conference, pp. 160–173. Boston, MA (1959)
3. Błażewicz, J., Lenstra, J., Rinnooy, A.: Scheduling subject to resource constraints: classification and complexity. Discret. Appl. Math. **5**, 11–24 (1983)
4. Brucker, P., Drexl, A., Möhring, R., Neumann, K., Pesch, E.: Resource-constrained project scheduling: notation, classification, models, and methods. Eur. J. Oper. Res. **112**, 3–41 (1999)

5. Demeulemeester E., Herroelen W.: Project scheduling: a research handbook. Kluwer Academic Publishers (2002)
6. Icmeli, O., Erenguc, S.S., Zappe, C.J.: Project scheduling problems: a survey. Int. J. Oper. Prod. Manag. **13**(11), 80–91 (1993)
7. Elmaghraby, S.E.: Activity nets: a guided tour through some recent developments. Eur. J. Oper. Res. **82**(3), 383–408 (1995)
8. Özdamar, L., Ulusoy, G.: A survey on the resource-constrained project scheduling problem. IIE Trans. **27**(5), 574–586 (1995)
9. Herroelen, W.S., De Reyck, B., Demeulemeester, E.L.: Resource-constrained project scheduling: a survey of recent developments. Comput. Oper. Res. **25**(4), 279–302 (1998)
10. Hartmann, S., Kölisch, R.: Experimental evaluation of state-of-the-art heuristics for the resource-constrained project scheduling problem. Eur. J. Oper. Res. **127**(2), 394–407 (2000)
11. Kölisch, R., Hartmann, S.: Experimental investigation of heuristics for resource-constrained project scheduling: an update. Eur. J. Oper. Res. **174**(1), 23–37 (2006)
12. Kölisch, R., Padman, R.: An integrated survey of deterministic project scheduling. OMEGA Int. J. Manag. Sci. **29**(3), 249–272 (2001)
13. Hartmann, S., Briskorn, D.: A survey of variants and extensions of the resource-constrained project scheduling problem. Eur. J. Oper. Res.**207**, 1–14 (2010)
14. Agarwal, A., Colak, S., Erenguc, S.: Metaheuristic methods. In: International Handbooks on Information Systems. Handbook on Project Management and Scheduling, vol. 1, pp. 57–74. (2015)
15. Alvarez-Valdés, R., Tamarit, J.M.: Heuristic algorithms for resource-constrained project scheduling: a review and empirical analysis. In: Słowiński, R., Węglarz, J. (eds.) Advances in Project Scheduling, Elsevier, Amsterdam, pp. 113–134 (1989)
16. Hartmann, S.: Project Scheduling under Limited Resources: Models, Methods, and Applications. Springer, Berlin Heidelberg (1999)
17. Knotts, G., Dror, M., Hartman, B.C.: Agent-based project scheduling. IIE Trans. **32**(5), 387–401 (2000)
18. Artigues, C., Demassey, S., Neron, E.: Resource-Constrained Project Scheduling: Models, Algorithms. Extensions and Applications. Robotics and Manufacturing Series. ISTE/Wiley, Control Systems (2008)
19. Brucker, P., Knust, S.: Complex Scheduling, 2nd edn. (2012)
20. Bowman, E.H.: The schedule-sequencing problem. Oper. Res. **7**, 621–624 (1959)
21. Brand, J.D., Meyer, W.L., Shaffer, L.R.: The Resource Scheduling Problem in Construction, Civil Engineering Studies, Report No 5. University of Illinois, Urbana, Department of Civil Engineering (1964)
22. Patterson, J.H., Roth, G.W.: Scheduling a project under multiple resource constraints: a zero-one programming approach. AIIE Trans. **8**, 449–455 (1976)
23. Deckro, R.F., Winkofsky, E.P., Hebert, J.E., Gagnon, R.: A decomposition approach to multi-project scheduling. Eur. J. Oper. Res. **51**, 110–118 (1991)
24. Icmeli, O., Rom: W.O.: Solving the resource-constrained project scheduling problem with optimization subroutine library. Comput. Oper. Res. **23**, 801–817 (1996)
25. Carruthers, J.A., Battersby, A.: Advances in critical path methods. Oper. Res. Q. **17**, 359–380 (1966)
26. Petroviç, R.: Optimisation of resource allocation in project planning. Oper. Res. **16**, 559–586 (1968)
27. Johnson, T.J.R.: An algorithm for the resource constrained project scheduling problem. PhD Dissertation, MIT (1967)
28. Schrage, L.: Solving resource-constrained network problems by implicit enumeration non-preemptive case. Oper. Res. **10**, 263–278 (1970)
29. Balas, E.: Project scheduling with resource constraints. In: Beale, E.M.L. (ed.) Applications of Mathematical Programming Techniques, pp. 187–200. American Elsevier, New York (1970)
30. Davis, E.W., Heidorn, G.E.: An algorithm for optimal project scheduling under multiple resource constraints. Manag. Sci. **27**, B803–B816 (1971)

31. Stinson, J.P., Davis, E.W., Khumawala, B.M.: Multiple resource-constrained scheduling using branch-and-bound. AIIE Trans. **10**(3), 252–259 (1978)
32. Talbot, B., Patterson, J.R.: An efficient integer programming algorithm with network cuts for solving resource-constrained scheduling problems. Manag. Sci. **24**, 1163–1174 (1978)
33. Radermacher, F.J.: Scheduling of project networks. Ann. Oper. Res. **4**, 227–252 (1985)
34. Christofides, N., Alvarez-Valdes, R., Tamarit, J.M.: Project scheduling with resource constraints: a branch and bound approach. Eur. J. Oper. Res. **29**, 262–273 (1987)
35. Bartusch, M., Mohring, R.H., Radermacher, F.J.: Scheduling project networks with resource constraints and time windows. Ann. Oper. Res. **16**, 201–240 (1988)
36. Bell, C.A., Park, K.: Solving resource-constrained project scheduling problems by A* search. Naval Res. Logist. **37**, 61–84 (1990)
37. Demeulemeester, E., Herroelen, W.: A branch-and-bound procedure for the multiple resource-constrained project scheduling problem. Manag. Sci. **38**, 1803–1818 (1992)
38. Demeulemeester, E.L., Herroelen, W.S.: New benchmark results for the resource constrained project scheduling problem. Manag. Sci. **43**(11), 1485–1492 (1997)
39. Carlier, J., Néron, E.: A new branch and bound method for solving the resource constrained project scheduling problem, PMS'96. In: The Fifth International Workshop on Project Management and Scheduling, pp. 61–65. Poznan, 11–13 April (1996)
40. Brucker, P., Knust, S., Schoo, A., Thiele, O.: A branch and bound algorithm for the resource-constrained project scheduling problem. Eur. J. Oper. Res. **107**(2), 272–288 (1998)
41. Mingozzi, A., Maniezzo, V., Ricciardelli, S., Bianco, L.: An exact algorithm for project scheduling with resource constraints based on a new mathematical formulation. Manag. Sci. **44**(5), 714–729 (1998)
42. Dorndorf, U., Pesch, E., Phan-Huy, T.: A branch-and-bound algorithm for the resource-constrained project scheduling problem. Math. Method Oper. Res. **52**, 413–439 (2000)
43. Kelley, J.: The critical-path method: resources planning and scheduling. In: Muth, J., Thompson, G. (eds.) Industrial scheduling, pp. 347–365. Prentice-Hall, New Jersey (1963)
44. Davis, E., Patterson, J.: A comparison of heuristic and optimum solutions in resource-constrained project scheduling. Manag. Sci. **21**, 944–955 (1975)
45. Cooper, D.F.: Heuristics for scheduling resource-constrained projects: an experimental investigation. Manag. Sci. **22**, 1186–1194 (1976)
46. Cooper, D.F.: A note on serial and parallel heuristics for resource-constrained project scheduling. Found. Control Eng. **2**, 131–133 (1977)
47. Boctor, F.F.: Some efficient multi-heuristic procedures for resource constrained project scheduling. Eur. J. Oper. Res. **49**, 3–13 (1990)
48. Özdamar, L., Ulusoy, G.: A local constraint based analysis approach to project scheduling under general resource constraints. Eur. J. Oper. Res. **79**, 287–298 (1994)
49. Ulusoy, G., Özdamar, L.: Heuristic performance and network/resource characteristics in resource-constrained project scheduling. J. Oper. Res. Soc. **40**, 1145–1152 (1989)
50. Thomas, P.R., Salhi, S.: A tabu search approach for the resource constrained project scheduling problem. J. Heuristics **4**, 123–139 (1998)
51. Leon, V.J., Ramamoorthy, B.: Strength and adaptability of problem-space based neighborhoods for resource-constrained scheduling. OR Spektrum **17**, 173–182 (1995)
52. Alcaraz, J., Maroto, C.: A robust genetic algorithm for resource allocation in project scheduling. Ann. Oper. Res. **102**, 83–109 (2001)
53. Alcaraz, J., Maroto, C., Ruiz, R.: Improving the performance of genetic algorithms for the RCPS problem. In: Proceedings of the Ninth International Workshop on Project Management and Scheduling, Nancy, pp. 40–43 (2004)
54. Hartmann, S.: A competitive genetic algorithm for resource-constrained project scheduling. Naval Res. Logist. **45**, 733–750 (1998)
55. Hartmann, S.: A self-adapting genetic algorithm for project scheduling under resource constraints. Naval Res. Logist. **49**, 433–448 (2002)
56. Coelho, J., Tavares, L.: Comparative analysis of metaheuricstics for the resource constrained project scheduling problem. Technical report, Department of Civil Engineering, Instituto Superior Tecnico, Portugal (2003)

57. Debels, D., Vanhoucke, M.: A decomposition-based genetic algorithm for the resource-constrained project scheduling problem. Oper. Res. **55**, 457–469 (2007)
58. Valls, V., Ballestín, F., Quintanilla, M.S.: A hybrid genetic algorithm for the RCPSP. Technical report, Department of Statistics and Operations Research, University of Valencia, (2003)
59. Valls, V., Ballestín, F.: A population-based approach to the resource-constrained project scheduling problem. Ann. Oper. Res. **131**, 305–324 (2004)
60. Agarwal, A., Colak, S., Erenguc, S.: A neurogenetic approach for the resource-constrained project scheduling problem. Comput. Oper. Res. **38**, 44–50 (2011)
61. Nonobe, K., Ibaraki, T.: Formulation and tabu search algorithm for the resource constrained project scheduling problem. In: Ribeiro, C.C., Hansen, P. (eds.) Essays and Surveys in Metaheuristics, pp. 557–588. Kluwer Academic Publishers, Springer Science+Business Media, New York (2002)
62. Artigues, C., Michelon, P., Reusser, S.: Insertion techniques for static and dynamic resource-constrained project scheduling. Eur. J.Oper. Res. **149**, 249–267 (2003)
63. Boctor, F.F.: An adaptation of the simulated annealing algorithm for solving resource-constrained project scheduling problems. Int. J. Prod. Res. **34**, 2335–2351 (1996)
64. Cho, J.H., Kim, Y.D.: A simulated annealing algorithm for resource-constrained project scheduling problems. J. Oper. Res. Soc. **48**, 736–744 (1997)
65. Bouleimen, K., Lecocq, H.: A new efficient simulated annealing algorithm for the resource-constrained project scheduling problem and its multiple modes version. Eur. J. Oper. Res. **149**, 268–281 (2003)
66. Valls, V., Ballestin, F., Quintanilla, M.S.: Justification and RCPSP: a technique that pays. Eur. J. Oper. Res. **165**, 375–386 (2005)
67. Zhang, H., Li, X.D., Li, H., Huang, F.L.: Particle swarm optimization-based schemes for resource-constrained project scheduling. Autom. Constr. **14**(3), 393–404 (2005)
68. Chen, R.M., Wu, C.L., Wang, C.M., Lo, S.T.: Using novel particle swarm optimization scheme to solve resource-constrained scheduling problem in PSPLIB. Expert Syst. Appl. **37**(3), 1899–1910 (2010)
69. Merkle, D., Middendorf, M., Schmeck, H.: Ant colony optimization for resource-constrained project scheduling. IEEE Trans. Evolut. Comput. **6**, 333–346 (2002)
70. Herbots, J., Herroelen, W., Leus, R.: Experimental investigation of the applicability of ant colony optimization algorithms for project scheduling. Research Report, KU, Leuven (2004)
71. Tseng, L.-Y., Chen, S.-C.: A hybrid metaheuristic for the resource-constrained project scheduling problem. Eur. J. Oper. Res. **175**, 707–721 (2006)
72. Akbari, R., Zeighami, V., Ziarati, K.: Artificial bee colony for resource constrained project scheduling problem. Int. J. Ind. Eng. Comput. **2**, 45–60 (2011)
73. Ziaratia, K., Akbaria, R., Zeighami, V.: On the performance of bee algorithms for resource-constrained project scheduling problem. Appl. Soft Comput. **11**(4), 3720–3733 (2011)
74. Jia, Q., Seao, Y.: Solving resource-constrained project scheduling problems: conceptual validation of FLP formulation and efficient permutation-based ABC computation. Comput. Oper. Res. **40**(8), 2037–2050 (2013)
75. Kochetov, Y., Stolyar, A.: Evolutionary local search with variable neighborhood for the resource constrained project scheduling problem. In: Proceedings of the 3rd International Workshop of Computer Science and Information Technologies, Russia (2003)
76. Debels, D., De Reyck, B., Leus, R., Vanhoucke, M.: A hybrid scatter search/electromagnetism meta-heuristic for project scheduling. Eur. J. Oper. Res. **169**(2), 638–653 (2006)
77. Paraskevopoulos, D.C., Tarantilis, C.D., Ioannou, G.: Solving project scheduling problems with resource constraints via an event list-based evolutionary algorithm. Expert Syst. Appl. **39**, 3983–3994 (2012)
78. Wang, L., Fang, C.: A hybrid estimation of distribution algorithm for solving the resource-constrained project scheduling problem. Expert Syst. Appl. **39**(3), 2451–2460 (2012)
79. Fang, C., Wang, L.: An effective shuffled frog-leaping algorithm for resource-constrained project scheduling problem. Comput. Oper. Res. **39**(5), 890–901 (2012)

80. Yannibelli, V., Amandi, A.: Hybridizing a multi-objective simulated annealing algorithm with a multi-objective evolutionary algorithm to solve a multi-objective project scheduling problem. Expert Syst. Appl. **40**(7), 2421–2434 (2013)
81. Sebt, M., Alipouri, Y.: Solving resource-constrained project scheduling problem with evolutionary programming. J. Oper. Res. Soc. **64**, 1327–1335 (2013)
82. Zamani, R.: A competitive magnet-based genetic algorithm for solving the resource-constrained project scheduling problem. Expert Syst. Appl. **229**(2), 552–559 (2013)
83. Fahmy, A., Hassan, T.M., Bassioni, H.: Improving RCPSP solutions quality with stacking justification-application with particle swarm optimization. Expert Syst. Appl. **41**(13), 5870–5881 (2014)
84. Patterson, J.H.: A comparison of exact approaches for solving the multiple constrained resource. Project scheduling problem. Manag. Sci. **30**(7), 854–867 (1984)
85. Kölisch, R., Sprecher, A., Drexl, A.: Characterization and generation of a general class of resource-constrained project scheduling problems. Manag. Sci. **41**, 1693–1703 (1995)
86. Ren, H., Wang, Y.: A survey of multi-agent methods for solving resource constrained project scheduling problems. In: Proceedings of International Conference on Management and Service Science, vol. 2011, pp. 1–4. (2011)
87. Shu-Guang, H., Er-Shi, Q., Gang, L.: A study on the project scheduling based on multi-agent systems. Math. Pract. Theory **1**, 43–47 (2005)
88. Wauters, T., Verbeeck, K., Berghe, G.V., De Causmaecker, P.: A multi-agent learning approach for the multi-mode resource-constrained project scheduling problem. In: Decker, S., Sierra, C. (eds.) Proceedings of 8th International Conference on Autonomous Agents and Multiagent Systems (AAMAS 2009), pp. 1–8. International Foundation for Autonomous Agents and Multiagent Systems. http://www.ifaamas.org
89. Wauters, T., Verbeeck, K., Berghe, G.V., De Causmaecker, P.: Learning agents for the multi-mode project scheduling problem. J. Oper. Res. Soc. **62**, 281–290 (2011)
90. Kölisch, R., Sprecher, A.: PSPLIB - A project scheduling problem library. Eur. J. Oper. Res. **96**, 205–216 (1996)
91. PSPLIB - Project Scheduling Problem LIBrary. http://www.om-db.wi.tum.de/psplib
92. Horenburg, T., Wimmer, J., Günthner, W.A.: Resource allocation in construction scheduling based on multi-agent negotiation. In: Proceedings of the 14th International Conference on Computing in Civil and Building Engineering (2012)
93. Zheng, X.-L., Wang, L.: A multi-agent optimization algorithm for resource constrained project scheduling problem. Expert Syst. Appl. **42**, 6039–6049 (2015)
94. Jędrzejowicz, P., Ratajczak-Ropel, E.: Agent-based approach to solving the resource constrained project scheduling problem. Lect. Notes Comput. Sci. **4431**, 480–487 (2007)
95. Laborie, P.: Complete MCS-based search: application to resource constrained project scheduling. In: Proceedings IJCAI-05, pp. 181–186. Edinburg, Scotland (2005)
96. Sprecher, A., Drexl, A.: Solving multi-mode resource-constrained project scheduling problems by a simple, general and powerful sequencing algorithm. Eur. J. Oper. Res. **107**, 431–450 (1998)
97. Jędrzejowicz, P., Ratajczak-Ropel, E.: New generation A-Team for solving the resource constrained project scheduling. In: Proceedings of the Eleventh International Workshop on Project Management and Scheduling, pp. 156–159. Istanbul (2008)
98. Jędrzejowicz, P., Ratajczak-Ropel, E.: Double-action agents solving the MRCPSP/Max problem. In: P. Jędrzejowicz et al. (eds.) Computational Collective Intelligence. Technologies and Applications. Lecture Notes in Artificial Intelligence, vol. 6923, pp. 311-321. (2011)
99. Jędrzejowicz, P., Ratajczak-Ropel, E.: Team of A-Teams for solving the resource-constrained project scheduling problem. In: Grana, M., Toro, C., Posada, J., Howlett, R., Lakhmi, C.J. (eds.) Advances in Knowledge Based and Intelligent Information and Engineering Systems. Frontiers in Artificial Intelligence and Applications, vol. 243, pp. 1201–1210. (2012)
100. Jędrzejowicz, P., Ratajczak-Ropel, E.: Reinforcement Learning Strategy for Solving the Resource-Constrained Project Scheduling Problem by a Team of A-Teams. In: Nguyen, N.T., Attachoo, B., Trawiński, B., Somboonviwat, K. (eds.) Intelligent Information and Database Systems. Lecture Notes in Artificial Intelligence, **8398**, 197–206. (2014)

101. Jędrzejowicz, P., Ratajczak-Ropel, E.: Reinforcement learning strategies for A-Team solving the resource-constrained project scheduling problem. Neurocomputing **146**, 301–307 (2014)
102. Glover F., Laguna M.: Tabu Search. Kluwer Academic Publishers (1997)
103. Glover, F.: Tabu search - Part I. ORSA J. Comput. **1**, 190–206 (1989)
104. Glover, F.: Tabu search - Part II. ORSA J. Comput. **2**, 4–32 (1989)
105. Glover, F.. Tabu search and adaptive memory programing: advances, applications and challenges. In: Barr, R.S., Helgason, R.V., Kennington, J.L. (eds.) Interfaces in Computer Science and Operations Research, pp. 1–75. Kluwer (1996)
106. Glover, F., Laguna, M., Marti, R.: Fundamentals of scatter search and path relinking. Control Cybern. **39**, 653–684 (2000)
107. Jędrzejowicz, P., Ratajczak-Ropel, E.: PLA based strategy for solving RCPSP by a team of agents. J. Univers. Comput. Sci. **22**(6), 856–873 (2016)
108. Jędrzejowicz, P., Ratajczak-Ropel, E.: Dynamic cooperative interaction strategy for solving RCPSP by a team of agents. In: Nguyen, N.T., Manolopoulos, Y., Iliadis, L., Trawiński, B. (eds.) Computational Collective Intelligence. Lecture Notes in Artificial Intelligence, vol. 9875, pp. 454–463. (2016)
109. Barto, A.G., Sutton, R.S., Anderson, C.W.: Neuronlike adaptive elements that can solve difficult learning control problems. IEEE Trans. Syst. Man Cybern. SMC-13, 835–846 (1983)
110. Sutton, R.S., Barto, A.G.: Reinforcement Learning: An Introduction. MIT Press, Cambridge, MA (1998)
111. Kaelbling, L.P., Littman, M.L., Moore, A.W.: Reinforcement learning: a survey. J. Artif. Intel. Res. **4**, 237–285 (1996)
112. Busoniu, L., Babuska, R., De Schutter, B.: A comprehensive survey of multiagent reinforcement learning. IEEE Trans. Syst. Man Cybern. Part C Appl. Rev. **38**(2), 156–172 (2008)
113. Tuyls, K., Weiss, G.: Multiagent learning: basics, challenges, prospects. AI Magazine **33**(3), 41–53 (2012)
114. Nareyek, A.: Choosing search heuristics by non-stationary reinforcement learning. In: Metaheuristics: Computer Decision-Making, pp. 523–544. Kluwer Academic Publishers (2001)
115. Barbati, M., Bruno, G., Genovese, A.: Applications of agent-based models for optimization problems: a literature review. Expert Syst. Appl. **39**, 6020–6028 (2012)
116. Jędrzejowicz, P.: Social learning algorithm as a tool for solving some difficult scheduling problems. Found. Comput. Decis. Sci. **24**(2), 51–66 (1999)
117. Palpant, M., Artigues, C., Michelon, P.: LSSPER: solving the resource-constrained project scheduling problem with large neighbourhood search. Ann. Oper. Res. **131**, 237–257 (2004)
118. Ranjbar, M.: Solving the resource-constrained project scheduling problem using filter-and-fun approach. Appl. Math. Comput. **201**, 313–318 (2008)
119. Fleszar, K., Hindi, K.: Solving the resource-constrained project scheduling problem by a variable neighbourhood search. Eur. J. Oper. Res. **155**, 402–413 (2004)
120. Valls, V., Ballestín, F., Quintanilla, S.: A hybrid genetic algorithm for the resource-constrained project scheduling problem. Eur. J. Oper. Res. **185**, 495–508 (2008)

Chapter 5
Multi-mode Resource-Constrained Project Scheduling

The first part of this chapter presents Multi-mode Resource-Constrained Project Scheduling Problem (MRCPSP) formulations and notations (Sect. 5.1). It also provides an overview of the best methods proposed so far for solving this problem, including a set of relevant bibliographic references in Sect. 5.2.

The second part of this chapter provides, in Sects. 5.3 and 5.4, an overview of agent-based approaches proposed for solving MRCPSP. This part includes methods, based on the A-Team approach, proposed by the author. Finally, in Sect. 5.4.6 the comparison of results is presented and some concluding remarks are drawn.

5.1 Problem Formulation

The Multi-mode Resource-Constrained Project Scheduling Problem (MRCPSP), introduced by Talbot in [1], is a generalization of the single-mode Resource-Constrained Project Scheduling Problem (RCPSP) described and formulated in Chap. 4. In case of the MRCPSP each activity j, $j = 0, \ldots, n+1$ may be executed in one out of M_j modes. The activities may not be preempted and a mode once selected may not change, i.e., a job j once started in mode m has to be completed in mode m without interruption. Performing job j in mode m takes d_{jm} periods and is supported by a set R^R of renewable and a set R^N of non-renewable resources. Hence, there are the set R^R of r^R renewable resource types and the set R^N of r^N non-renewable resource types considered. The availability of each renewable resource type $k \in R^R$ in each time period is r_k^R units, $k = 1, \ldots, r^R$. Each mode m of activity j requires r_{jmk}^R units of renewable resource k during each period of its duration, where $r_{01k}^R = r_{n+11k}^R = 0$, $k = 1, \ldots, r^R$. The availability of each non-renewable

E. Ratajczak-Ropel and A. Skakovski, *Population-Based Approaches to the Resource-Constrained and Discrete-Continuous Scheduling*,
Studies in Systems, Decision and Control 108, DOI 10.1007/978-3-319-62893-6_5

resource type l is r_l^N units in total, $l = 1, \ldots, r^N$. Each mode m of activity j requires r_{jml}^N units of resource l during its duration, where $r_{01l}^N = r_{n+11l}^N = 0, l = 1, \ldots, r^N$.

The structure of a project, as in the RCPSP, can be represented by an Activity On Node (AON) network $G = (V, E)$, where V denotes the set of activities and E is the set of precedence relationships between these activities. S_j (P_j) is the set of successors (predecessors) of activity j, $j = 0, \ldots, n+1$. It is further assumed that $0 \in P_j, j = 1, \ldots, n+1$, and $n+1 \in S_j$, $j = 0, \ldots, n$.

The objective, as in the RCPSP case, is to find the schedule S of activities starting times $[s_1, \ldots, s_n]$, where $s_1 = 0$, the precedence and resource-constraints are satisfied, such that the project complexion time (schedule duration, makespan) $CT(S) = s_{n+1}$ is minimized.

The mathematical model of the MRCPSP was introduced by Talbot [1]. Is was based on the Pritsker's RCPSP model presented in Sect. 4.1. Binary variable $x_{jmt} = 1$ if activity j executed in mode m is completed at the end of time period t, otherwise, $x_{jmt} = 0$. With each activity j its earliest and latest finish time is associated, denoted respectively EF_j and LF_j. Both are calculated assuming that the shortest duration mode is assigned to each activity. The planning horizon H is calculated for the modes with the longest durations, it never exceeds the sum of all activity durations.

The following formulas (5.1)–(5.6) define the problem:

$$\min \sum_{t=EF_{n+1}}^{LF_{n+1}} t x_{n+1,m,t} \tag{5.1}$$

$$\sum_{m=1}^{|M_j|} \sum_{t=EF_j}^{LF_j} x_{jmt} = 1, \quad \text{for } j = 0, \ldots, n+1 \tag{5.2}$$

$$\sum_{m=1}^{|M_j|} \sum_{t=EF_i}^{LF_i} t x_{imt} \leq \sum_{m=1}^{|M_j|} \sum_{t=EF_j}^{LF_j} t x_{jmt} - d_{jm}, \quad \text{for all } (i,j) \in P \tag{5.3}$$

$$\sum_{j=1}^{n} \sum_{m=1}^{|M_j|} \sum_{q=max\{t,EF_j\}}^{min\{t+d_{jm}-1,LF_j\}} r_{jmk}^R x_{jmq} \leq r_k^R, \text{ for } k = 1, \ldots r^R, \ t = 1, \ldots H \tag{5.4}$$

$$\sum_{j=1}^{n} \sum_{m=1}^{|M_j|} \sum_{t=EF_j}^{LF_j} r_{jml}^N x_{jmt} \leq r_l^N, \text{ for } l = 1, \ldots r^N \tag{5.5}$$

$$x_{jmt} \in \{0, 1\}, \text{ for } i = 0, \ldots, n+1, \ m \in M_j, \ t = EF_j, \ldots, LF_j \tag{5.6}$$

Constraint (5.2) ensure that each activity is performed exactly once in exactly one mode. Precedence constraints are guaranteed by (5.3). Constraints (5.4) and (5.5) guarantee that the renewable and non-renewable resource limits are not exceeded, respectively. Constraints (5.6) assures that binary decision variables are considered.

The MRCPSP is NP-hard in the strong sense as a generalization of the RCPSP (see Błażewicz et al. [2]). Moreover, if there is more than one nonrenewable resource, the problem of finding a feasible solution for the MRCPSP is NP-complete [3]. The considered problem class is denoted as $MPS|prec|C_{max}$ according to the classification scheme of Brucker et al. [4] or it is denoted as $m, 1T|cpm, disk, mu|C_{max}$ according to the classification scheme of Demeulemeester and Herroelen [5].

The project instance of MRCPSP, as in the RCPSP, is usually represented by an Activity On Node (AON) network. Sequences of activities are build using the priority rules defined for the RCPSP. To generate a schedule from the sequence, the serial, parallel or combined SGS is used, and for determining the duration and the resource requirements of the activities a mode priority/selection rule is necessary. The different mode selection rules were proposed, among others, in [6, 7]. The examples of mode selection rules are as follow:

- Shortest Feasible Mode (SFM),
- Least Product Sum of Resource and Duration (LPSRD),
- Least Total Resource Usage (LTRU),
- Least Critical Resource Usage (LCRU),
- Least Resource Proportion (LRP),
- Least Sum of non-renewable Resource (LRS).

Three methods of placing activity modes in the schedule, as in the RCPSP, are used: forward, backward and forward-backward. To improve a solution after applying the backward scheduling, Local Left Shift procedure (LLS) is usually used.

5.2 State of the Art Review

In this section state of the art for the MRCPSP formulated in Sect. 5.1 is reviewed. The methods for solving MRCPSP are, as in case of the RCPSP, divided into three groups: exact, heuristic and metaheuristic ones. The MRCPSP is a generalization of RCPSP, hence in many papers both problems are considered together.

An overview of the exact methods for the MRCPSP was given by Sprecher in [8]. Exact methods based on branch-and-bound algorithms were investigated by Hartmann and Drexl [9]. A comprehensive survey of the RCPSP and MRCPSP is presented by Herroelen et al. [10], Brucker et al. [4], Kölisch and Padman [11] and Demeulemeester and Herroelen [5]. A recent literature review on project scheduling problem variants is provided by Hartamnn and Briskorn [12]. Other group of papers deal with the MRCPSP only. A recent survey of project scheduling with finite or infinite number of activity processing modes is provided by Węglarz et al. [13]. An overview and experimental investigation of the existing metaheuristic solution procedures to solve the MRCPSP can be found in Peteghem and Vanhoucke [14]. Authors compared over 20 existing approaches. A recent overview and state of the art of the MRCPSP with different objectives can be found in the paper of Mika et al. [15].

The exact methods applied to the MRCPSP, similarly as for the RCPSP, can be classified into mathematical programming i.e. zero-one programming and implicit enumeration including dynamic programming and branch-and-bound (B&B). The first solution method for the MRCPSP was linear programming approach proposed by Słowiński [16]. Other examples of research involving the use of mathematical programming can be found in the papers of Talbot [1] and Patterson et al. [17]. In these papers the enumeration scheme-based algorithm for the MRCPSP was proposed. Many authors considered brunch-and-bound based approaches, for example: Speranza and Vercellis [18], Hartmann and Sprecher [19], Sprecher et al. [8, 20], Hartmann and Drexl [9], and Sprecher and Drexl [21]. Branch-and-cut algorithm was developed by Zhu et al. [22].

Considering the MRCPSP computational complexity, the problem, like in case of the RCPSP, can be solved by exact methods only for a small size problem instances. For more complex problem sets, which are common in practice, solution techniques have to be focused on heuristic and metaheuristic methods, if a solution is required to be computed in a reasonable time.

Heuristic solution procedures for the MRCPSP are generally based on the priority rules. Talbot [1] proposed the set of priority rules and used them in his B&B algorithm. The other set of priority rules was proposed by Boctor [6]. Drexl and Grünewald [23] proposed the stochastic scheduling method named STOCOM. Heuristic approach called local constraint based analysis (LCBA) was proposed by Özdamar and Ulusoy [24]. In [25] Özdamar proposed several priority rules and used them in genetic algorithm. Kölisch and Drexl [26] proposed a local search based method and compared the results with two other heuristics.

Many metaheuristic methods have been applied to solve the MRCPSP. The most common used are: genetic algorithms (GA), scatter search (SS), simulated annealing (SA), particle swarm optimization (PSO), ant colony optimization (ACO), differential evolution algorithm (DEA), estimation of distribution algorithms (EOD), multi-agent learning algorithms (MALA) described in Sect. 5.3, and non-standard methods combining different approaches.

Many authors proposed the efficient approaches based on the genetic algorithms (GA). Example papers include: Mori and Tseng [27], Özdamar [28], Hartmann [29], Alcaraz et al. [30], Tseng and Chen [31], Peteghem and Vanhoucke [32], and Coelho and Vanhoucke [33]. Hybrid genetic algorithm that uses a local search method to improve solutions generated by the GA was proposed by Lova et al. [34]. Elloumi and Fortemps developed hybrid rank-based EA for the MRCPSP. Scatter search (SS) based approaches to the MRCPSP described Ranjbar et al. [35] and Peteghem and Vanhoucke [36]. Simulated annealing (SA) based approaches were proposed by Słowiński et al. [37], Boctor [7], Józefowska et al. [38] or Bouleimen and Lecocq [39]. Approaches based on PSO were developed by Zhang et al. [40] and Jarboui et al. [41]. Ant colony optimization (ACO) based approaches proposed Chiang et al.

[42]. Differential evolution algorithm (DEA) developed Damak et al. [43]. Tchao and Martins [44] described some hybrid heuristics based on TS and PR. Wang and Fang [45] proposed shuffled frog-leaping algorithm (SFLA) for solving the MRCPSP.

Recent works on the MRCPSP includes exact approaches as well as metaheuristics. Due to the fact that the MRCPSP is one of the most intractable and challenging problems in the Operational Research, newly proposed optimization techniques are usually adapted to solve it. They include swarm optimization, population based and hybrid methods as well as agent based approaches.

Kyriakidis et al. [46] proposed mixed-integer linear programming (MILP) formulations for the RCPSP and MRCPSP based on the resource-task network (RTN) representation and solved them using the proposed formulations. Wang and Fang in [47] proposed estimation of distribution algorithms (EOD). Li and Zhang [48] proposed ant ACO based methodology for solving the MRCPSP and compared it with existing metaheuristics. Genetic algorithm based on random key and related mode assignment representation scheme was proposed by Sebt et al. [49]. The above authors also compared the results with earlier algorithms. Geiger [50] developed a multi-threaded local search algorithm for the Multi-mode Resource-Constrained Multi-Project Scheduling Problem (MRCMPSP) based on the concepts of variable neighborhood search (VNS), together with iterated local search (ILS) and proposed the parallel (multi-core) implementation for it. Author tested the algorithm on MRCPSP using MMLIB [51] benchmark data sets and compared the results with others reported in the literature.

Most of the optimal, heuristic and metaheuristic methods for the MRCPSP were tested on two well-known benchmark data sets, the PSPLIB dataset [52, 53] and the Boctor dataset [6]. Recently Peteghem and Vanhoucke in [14] have proposed new benchmark data sets MMLIB [51]. In the paper authors indicates some shortcomings of the PSPLIB data sets which could lead to biased results. Geiger [54] has proposed a checker software for the MRCPSP solution files. The program supports, in particular, the problem instances from MMLIB.

A large number of project data for integrated project management and control, including artificial and real data sets, can be found on the web page of Operations Research and Scheduling research group [55]. There are also two network generators to construct project networks under a controlled design accessible to download, based on the principles proposed by Demeulemeester et al. [56].

5.3 Agent-Based Approaches to MRCPSP

A few MAS implementations exist for the Multi-mode Resource-Constrained Project Scheduling Problem (MRCPSP). All except one were validated experimentally using the data sets from PSPLIB.

Knotts et al. in [57, 58] proposed MAS model based on standard electronic components used in a simple digital circuits. An agent's precedence and resource constraints are represented as a combination of AND, OR, NAND, NOR and XOR gates.

For each activity of the project one agent is created which is responsible for acquiring the resources required by this activity. Two types of agents, differing in their behavior, are considered: basic and enhanced. The basic agent is purely reactive, it becomes active in the first feasible execution mode that it finds. The enhanced agent is capable of deliberative behavior, it deliberates in regard to mode selection according to several rules. The system consists of the set of agents, blackboard and three modules: problem generator module, simulation manager module and activity duration realizer module. The globally accessible blackboard stores all information about the current state of the system including the precedence and resource requirements of activities. Agents scan the blackboard to determine whether or not their precedence and resource requirements can be satisfied, remove the resources required during their duration and return renewable resources when their duration passed.

For the algorithms proposed by Knotts et al. [57, 58] the stochastic simulation experiment was carried out using 500 problem instances generated from the project network including 51 activities, originally proposed by Maroto and Tormos [59]. In the experiment two types of agents, as well as, eight different priority rules: SPT, LPT, FIS, MIS, SRR, GRR, EST and EDD, used to control agent access to resources, have been tested. Simulation showed that the results obtained by the approach are comparable with the results known from the literature and commercial systems. The proposed algorithms could solve large-scale MRCPSP instances. In the above mentioned papers one can find also performance comparison and analysis of multi-agent systems with two types of agents and different priority rules. Using enhanced agents gives better results than using basic agents. The SPT happened to be the best priority rule when applied by the basic agents. The EDD occurred the best priority rule when applied by the enhanced agents.

Wauters et al. [60, 61] implemented and used a network of distributed reinforcement learning agents that cooperate to jointly learn a well-performing constructive heuristic. The approach is based on an Activity On Node (AON) network, which is usually used to represent precedence relation between the activities in RCPSP and MRCPSP. An agent is placed in each activity node and one extra dispatching agent (dispatcher) is added. The main idea of the algorithm is to enable each agent to learn in which order to visit its successors, and in which mode the activity needs to be performed. In each step of the algorithm one of agents has a control and makes decision. This agent chooses an order to visit its successors and a mode and passes the control to the first agent from this order. This process is continued until the agent in the last dummy node gives back the control. Then, it forwards the control to the dispatcher, which chooses a random eligible agent from the list of already visited ones or a random eligible unvisited agent. These procedure is repeated until all agents have been visited at least once. Moreover, at any time agent can give the control, with a small probability, to the dispatcher. As a result the activity list (AL) is obtained. To construct the schedule from the AL the serial SGS is used. For learning the activity order and the best modes, reinforcement learning is used as learning automata with linear reward-inaction algorithm.

The system of Wauters et al. [61] was implemented in Java and validated experimentally. To validate the approach the datasets from PSPLIB have been used, for the MRCPSP instances from j10, j12, j14, j16, j18, j20 and j30 data sets. The approach was additionally tested for the single-mode RCPSP dataset j120 including instances with 120 activities. In both cases the experiments show that the obtained results are comparable with the others results known from the literature and can be useful for solving the considered problems instances.

Mirzaei and Akbarzadeh in [62] proposed Multi-Agent Learning Algorithm (MALA) for solving the MRCPSP. The approach is based on the AON network where activities on the project are considered as agents. For each agent two devices for making its decisions were implemented: learning atomaton and heuristic-based stochastic local dispatcher. The approach is similar to the described above Wauters et al. [61] approach called in this paper MARLA. The global dispatcher was used in exactly the same way. Each agent uses two learning automata to choose its own execution mode and also the order of visiting its successor activities. In comparison with MARLA [61] authors indicate two main differences. Firstly, the schedules are constructed locally. Secondly, local heuristic-based stochastic dispatcher have been added to each agent as its second tool for making decisions. As a result the activity list (AL) is obtained. To construct the schedule from the AL the serial SGS is used.

The algorithm of Mirzaei and Akbarzadeh [62] was implemented in MATLAB and validated experimentally using the MRCPSP instances from PSPLIB. In the first experiment the average computation times of two agent systems MALA and MARLA and two metaheuristic algorithms from the literature (PSO and GA) have been compared using nine instances from different PSPLIB datasets. In the second experiment the performance of the same algorithms has been compared for three PSPLIB data sets: j10, j16 and j20. The experiments show the effectiveness of MALA but the differences in performance are very small. It is worth noticing that both described agent based approaches MALA and MARLA requires less computational time than the other two heuristic methods.

Wenzler and Günthner [63] proposed the MAS for solving the MRCPSP similar to Wauters et al. [61] approach. Different types of agents represent activities (process agents), renewable and nonrenewable resources. The central blackboard is used to negotiate the resource allocation to current activities. All available resources are registered on the blackboard. All activity modes register themselves on the blackboard with resource requests and calculated priority value as soon as their predecessors finish. Agent which activity finishes informs all the former's successors about the completion. Additionally, the learning agent is used to analyze the previous planning actions, influence the process agents' mode choices, restarts the planning procedure and compares the result with the previous solution. The learning agent's behavior is controlled by heuristic. Its two main tasks are creating a feasible solution and improving the feasible solution as far as possible.

The MAS proposed by Wenzler and Günthner was implemented as a discrete event simulation (DES) system. The preliminary computational experiment has been carried out using datasets from PSPLIB. Firstly, the effectiveness of the proposed MAS in comparison with the same MAS without learning agent was shown.

Secondly, the results obtained for datasets j10, j12, j14, j16, j18, j20 and j30 were compared with the best alternative methods known from the literature. The preliminary results are worse but some modifications are proposed.

5.4 A-Teams Solving the MRCPSP

In this section five variants of A-Teams implemented for solving the MRCPSP are presented. All A-Teams have been constructed using JABAT environment. For each A-Team variant a computational experiment has been carried out and results have been discussed. In Sect. 5.4.1 the two first approaches with the single A-Teams using the static cooperation strategy have been described, as well as the details of the computational experiments which are the same for all of them. In Sect. 5.4.2 the essential modifications of the first approaches are indicated and the implementations of algorithms used in the next A-Teams are presented. The next three sections include descriptions of the single A-Teams with dynamic cooperation strategies based on: reinforcement learning (RL) in Sect. 5.4.3, Population Learning in Sect. 5.4.4, and integration of the best rules from previous approaches in Sect. 5.4.5. In each case the computational experiments results follow. Finally in Sect. 5.4.6 the summary of the proposed approaches are presented including comparison of the results with the best results known from the literature.

5.4.1 Single A-Teams with the Static Cooperation Strategies

First A-Team based approaches to the RCPSP and MRCPSP were proposed in [64, 65]. Base classes representing the RCPSP and MRCPSP instances were implemented using JABAT middleware. The RCPSP variants are introduced in Sect. 4.4. In order to solve the MRCPSP instances, classes representing activity and resource have been adopted to store additional data and a new class mmMode is implemented identifying the activity mode. Therefore, the proposed approach includes implementations of the following classes representing the MRCPSP: mmData, mmTask, mmSolution, mmActivity, mmResource, and mmMode. The mmResource class identifies both - renewable and non-renewable resource, storing the value representing the number of the resource units. The mmMode class identifies activity mode, which attributes include the mode number, duration and two lists of the required resources of both types. The mmActivity class, instead of activity duration and list of resources, stores a list of modes representing by the mmMode class.

Six optimization algorithms are implemented and used as optimization agents. These are as follows:

- Local Search Algorithm (mmLSA),
- Tabu Search Algorithm (mmTSA),

- Crossover Algorithm (mmCA),
- Minimal Critical Sets based Algorithm (mmMCSA),
- Precedence Tree Algorithm (mmPTA),
- Path Relinking Algorithm (mmPRA).

Each of them is adopted for the MRCPSP to work with the set of new classes. Additionally, multiple modes of each activity and two types of resources are considered. For example, mmCA additionally check all possible modes of each activity, and in mmLSA and mmTSA the move is adopted to represent also mode exchange.

Because the discussed algorithms have been undergoing several changes over various stages of development, final detailed versions used in the following computational experiment are described further on in Sect. 5.4.2.

All optimization agents work together and interact with the common memory according to the basic interaction strategy, the same as in case of the RCPSP:

- The initial population is generated randomly,
- The individuals from the population are chosen randomly and immediately send to optimization agents,
- An improved solution replaces the worst one from the population,
- The computation for one problem instance in this approach is interrupted after 5 min.

The approach proposed for the MRCPSP differs in the set of optimization agents used. In these approaches the following optimization algorithms are implemented and used as optimization agents:

- OAmmTSAe - implementing the Tabu Search Algorithm mmTSA,
- OAmmCA - implementing the Crossover Algorithm mmCA,
- OAmmMCSA - implementing the Minimal Critical Sets based Algorithm mmMCSA
- OAmmPTA - implementing the Precedence Tree Algorithm mmPTA,
- OAmmPRA - implementing the Path Relinking Algorithm mmPRA.

The first experiments have been performed for the single A-Team with the basic static cooperation strategy (SC Strategy) described above using OAmmLSA, OAmmTSA, OAmmCA, OAmmMCSA and OAmmPTA. In the following part of this work the approach is called SC1 Strategy.

The second series of experiments differs in the set of optimization agents and cooperation strategy elements. In the set of optimization agents instead of the least effective OAmmMCSA the OAmmPRA was used. In the cooperation strategy an improved solution replaces the original one instead of the worse one from the population (common memory).

Table 5.1 PSPLIB test sets of the MRCPSP

Number of activities	10	12	14	16	18	20
Number of instances	536	547	551	550	552	554

The second set of experiments performed for the single A-Team with the static cooperation strategy (SC Strategy) using OAmmLSA, OAmmTSA, OAmmCA, OAmmPTA and OAmmPRA is called SC2 Strategy.

The proposed approaches have been validated experimentally using benchmark instances test sets of the MRCPSP from PSPLIB. The following mm (multi mode) test sets are used: j10 (multi-mode, 10 activities), j12, j14, j16, j18, and j20. For each set 640 problem instances was generated, however the infeasible ones were removed [53, 66]. Hence, the instances for which at least one feasible solution has been found are used. The actual numbers of MRCPSP instances in considered data sets are shown in Table 5.1.

The experiment involved computation with a fixed number of optimization agents, fixed population sizes, and fixed stopping criteria.

During all experiments the following characteristics of the computational results have been calculated and recorded: Mean Relative Error (Mean RE) and Maximal Relative Error (Max RE) calculated as the deviation from the optimal solution and the Critical Path Lower Bound (CPLB), Mean Computation Time (Mean CT) which has been needed to find the best solution and Mean Total Computation Time (Mean TCT) which has been needed to stop all optimization agents and the whole system. Each instance has been solved five times and the results have been averaged over these solutions.

All experiments have been carried out using nodes of the cluster Holk of the Tricity Academic Computer Network built of 256 Intel Itanium 2 Dual Core 1.4 GHz with 12 MB L3 cache processors and with Mellanox InfiniBand interconnections with 10 Gb/s bandwidth.

The results presented in this book are recalculated in respect of the number of generated schedules (SGS procedure calls, see Sect. 5.1), which has been limited to 5000. Some of the algorithms implementations have been modified and improved in respect of the Java language structures.

The results obtained for the two first approaches to solve the MRCPSP by the A-Team implemented in JABAT are summarized in Tables 5.2 and 5.3. It can be seen that the second approach is most effective with respect to the Mean RE and Mean CT. The best improvement was obtained for the set j18 including instances with 18 activities.

Table 5.2 Experiment results for the A-Team with SC1 Strategy solving the MRCPSP

Number of activities	Mean RE from opt*/bk solution [%]	Max RE from opt*/bk solution [%]	Mean RE from CPLB [%]	Max RE from CPLB [%]	Mean CT [s]	Mean TCT [s]
10	0.75	10.53	30.77	314.29	43	76
12	0.76	12.5	25.42	166.67	51	88
14	0.80	25.0	22.57	193.33	51	97
16	0.83	11.11	19.00	164.29	50	91
18	0.98	18.00	18.68	153.33	40	76
20	1.85	22.58	18.60	161.54	59	115

Table 5.3 Experiment results for the A-Team with SC2 Strategy solving the MRCPSP

Number of activities	Mean RE from opt*/bk solution [%]	Max RE from opt*/bk solution [%]	Mean RE from CPLB [%]	Max RE from CPLB [%]	Mean CT [s]	Mean TCT [s]
10	0.69	18.75	30.66	314.29	22	46
12	0.70	23.53	25.72	207.14	29	59
14	0.76	18.75	22.62	193.33	30	63
16	0.79	10.79	19.09	164.29	25	47
18	0.91	18.75	18.65	153.33	23	40
20	1.81	21.54	18.50	156.16	29	64

5.4.2 *Algorithms Used in the Further A-Team Approaches*

During the first experiments some problems and disadvantages of the environment, strategies and optimization agents were occurred which resulted in several modifications. In this section the algorithms used in all subsequent JABAT based A-Team approaches are presented.

To describe the above algorithms and present their pseudocodes the following auxiliary functions are used:

- minCT - requires two schedules (individuals) as arguments and returns the better one with respect to the project duration (see Fig. 4.2 in Sect. 4.4.1).
- mmMakeMove - moves the activity form one position in the schedule (activity list) to the another position and simultaneously changes its active mode, the modified schedule is returned as the result (Fig. 5.1).
- mmReverseMove - cancels the move in the schedule and returns the previous schedule as the result (Fig. 5.2)
- mmMakeExchange - exchanges the activities at the two given positions in the schedule and simultaneously changes its active mode, as the result returns the modified schedule (Fig. 5.3).

```
mmMakeMove(S, pi, m_pi, m_pi^new, pj)
{
    S0 = S
    Change active mode of activity from position pi from m_pi to m_pi^new
    Move activity from position pi to position pj in S
    if(precedence and resource constraints in S are satisfied)
        return S
    else return S0
}
```

Fig. 5.1 Pseudocode of the mmMakeMove function

```
mmReverseMove(S, pi, m_pi, m_pi^new, pj)
{
    Move activity from position pj to position pi in S
    Change active mode of activity from position pi from m_pi^new to m_pi
    return S
}
```

Fig. 5.2 Pseudocode of the mmReverseMove function

```
mmMakeExchange(S, pi, m_pi, m_pi^new, pj, m_pj, m_pj^new)
{
    S0 = S
    Change active mode of activity from position pi from m_pi to m_pi^new
    Change active mode of activity from position pj from m_pj to m_pj^new
    Exchange activities from positions pi and pj
    if(precedence and resource constraints in S are satisfied)
        return S
    else return S0
}
```

Fig. 5.3 Pseudocode of the mmMakeExchange function

- mmReverseExchange - reverses the exchange move in the schedule and returns the previous schedule as the result (Fig. 5.4).

Active mode is understood as the current mode used to construct the schedule.

The pseudocodes of the optimization algorithms are presented in Figs. 5.5, 5.6, 5.7 and 5.8.

The mmLSA is a local search algorithm which finds local optimum by moving each activity to all possible places in the schedule (in mmLSAm) or by exchanging pairs of activities (in mmLSAe), simultaneously changing the active modes of these activities. After such defined moves of the local search the emerging solution (i.e. schedule) is generated through applying SGS. For each combination of activities and modes the value of possible solution is calculated. The best schedule is remem-

```
mmReverseExchange(S, pi, m_pi, m_pi^new, pj, m_pj, m_pj^new)
{
    Exchange activities from positions pi and pj in S
    Change active mode of activity from position pi from m_pi^new to m_pi
    Change active mode of activity from position pj from m_pj^new to m_pj
    return S
}
```

Fig. 5.4 Pseudocode of the mmReverseExchange function

```
mmLSAm(S, itNumber, iStep, fStep)
{
    bestS = S
    it = 0
    while(it < itNumber)
    {
        for(step = iStep; step < fStep; step = step + 1)
        {
            pi = ++pi % (n - step) + 1
            for(pj = pi; pj < n - step; pj = pj + step)
            {
                m_pi = active mode of activity from position pi
                for(each mode m_pi^new of activity from position pi except m_pi)
                {
                    mmMakeMove(S, pi, m_pi, m_pi^new, pj)
                    SGS(S)
                    if(CT(S) < CT(bestS)) bestS = S
                    mmReverseMove(S, pi, m_pi, m_pi^new, pj)
                }
            }
        }
        it++
    }
    return bestS
}
```

Fig. 5.5 Pseudocode of the mmLSAm algorithm

bered and finally returned. The pseudocode of the mmLSAm is shown in the Fig. 5.5. The mmLSAm procedure requires four variables, just like the smLSAm. The first, S denotes an initial schedule (individual). The second, *itNumber* is the number of iteration for the algorithm. The last two, *iStep* and *fStep* indicate the initial and final distance between activities under moving or exchange. The pseudocode of the mmLSAe algorithm differs mainly in using the mmMakeExchange instead of mmMakeMove function and mmReverseExchange instead of mmReverseMove function, hence it is not presented separately.

```
mmTSAe(S, pi, maxItNWI, step)
{
    TabuList = ∅
    it = 0
    bestS = S
    while(it < maxItNWI)
    {
        bestSit = null
        for(pj = pi+1;  pj < n−1;  pj = pj+step)
        {
            m_pi = active mode of activity from position pi
            m_pj = active mode of activity from position pj
            for(each mode m_pi^new of activity from position pi)
                for(each mode m_pj^new of activity from position pj)
                    if(m_pi^new != m_pi || m_pj != m_pj^new)
                    {
                        move = (S, pi, m_pi, m_pi^new, pj, m_pj, m_pj^new)
                        if(move is not in TabuList)
                        {
                            mmMakeExchange(move)
                            SGS(S)
                            if(CT(S) < CT(bestSit)) bestSit = S
                            mmReverseExchange(move)
                        }
                    }
        }
        Update TabuList
        if(bestSit is not null)
        {
            if(T(bestSit) < T(bestS))
            {
                bestS = bestSit
                it = 0
            }
            add moves to TabuList:
                (pi, mpi, mi, pj, mpj, mj, maxItNWI)
                (null, null, null, pi, mpi, null, ⌊n/4⌋)
                (null, null, null, pi, mi, null, ⌊n/4⌋)
                (pi, mpi, null, null, null, null, ⌊n/2⌋)
                (pi, mi, null, null, null, null, ⌊n/2⌋)
                (pj, null, null, pi, null, null, ⌊maxItNWI/2⌋)
        }
        else it++
        pi = pi % (n−2) + 1
    }
    return bestS
}
```

Fig. 5.6 Pseudocode of the mmTSAe algorithm

```
mmCA(S1, S2, step)
{
    bestS = minCT(S1,S2);
    for (pi = 1;  pi ≤ n;  pi = pi + step)
    {
        Do simple (one point) crossover on S1 and S2 in point pi
        m_pi = active mode of activity from position pi
        for (each mode m_pi of activity from position pi)
        {
            Change active mode m_pi of activity from position pi to m_pi^new in S1
            SGS(S1)
            Change active mode m_pi of activity from position pi to m_pi^new in S2
            SGS(S2)
            bestS = minCT(bestS, minCT(S1,S2));
        }
    }
    return bestS;
}
```

Fig. 5.7 Pseudocode of the mmCA algorithm

```
mmPRA(S1, S2)
{
    bestS = minCT(S1,S2);
    S = S1
    for(pi = 1;  pi ≤ n;  pi++)
    {
        m_pi =  active mode of activity from position  pi
        for (each mode m_pi^new of activity from position pi except m_pi)
        {
            pj = position in S of activity from position pi in S2
            mmMakeMove(S, pi, m_pi, m_pi^new, pj)
            if(CT(S) < CT(bestS))  bestS = S
        }
    }
    return bestS
}
```

Fig. 5.8 Pseudo-codes of the mmPRA algorithm

The mmTSAe algorithm is an implementation of tabu search metaheuristic [67–69]. It finds local optimum by exchanging each two activities in the schedule, for which such exchange is possible. The best exchange move from the neighborhood of the solution, which is not tabu, is chosen and performed. The pseudocode of the proposed mmTSAe is shown in Fig. 5.6. The algorithm smTSAe was adjusted to solving the MRCPSP. The main differences are using the modes and checking different modes of exchanged activities. For this reason the new exchange move was defined including the active (current) modes and new modes of both exchanged activities.

The tabu list *TabuList* has been implemented to store the information of recent moves which should not be repeated by a fixed number of iterations. For example:

- Setting the exchange move ($pi, mpi, mi, pj, mpj, mj, maxItNWI$) as the tabu one prevents from repeating the same exchange by *maxItNWI* iterations. It blocks for *maxItNWI* iterations all moves that exchange activity from position *pi* with activity from position *pj* and simultaneously change the active mode of activity from position *pi* from *mpi* to *mi*, and active mode of activity from position *pj* from *mpj* to *mj*.
- Setting the exchange move (null, null, null, pi, mpi, null, $\lfloor n/4 \rfloor$) as the tabu one blocks exchanges of any activity with activity from position *pi* with active mode *mpi* for $\lfloor n/4 \rfloor$ iterations.

The best schedule is remembered and finally returned.

The pseudocode of the mmTSAm differs mainly in using the mmMakeMove instead of mmMakeExchange function and the mmReverseMove instead of mmReverseExchange function, so it is not presented separately.

The mmCA is an algorithm based on the idea of the one point crossover operator. For a pair of solutions $S1$, $S2$ one point crossover is applied. The parameter *step* determines frequency this operation is performed. Next, for the crossover offspring at point *pi* both solutions are evaluated for different new modes. The best schedule is remembered and finally returned. The pseudocode of the mmCA is shown in the Fig. 5.7.

The mmPRA, as in case of the RCPSP, is the implementation of the path-relinking algorithm [70, 71]. For a pair of solutions a path between them is constructed. The path consists of schedules obtained by carrying out a single move from the preceding schedule. Such move is performed by the mmMakeMove function (see Fig. 5.1). For each schedule in the path the value of the respective solution is checked. The best schedule is remembered and finally returned. The pseudocode of the mmPRA is shown in the Fig. 5.8.

Four optimization algorithms originally proposed in [72] are used in all subsequent A-Teams implementations described in this section. There are: mmLSAm, mmTSAe, mmCA, and mmPRA (see Figs. 5.5, 5.6, 5.7 and 5.8). The following optimization agents derived from OptiAgent class are used:

- OAmmLSAm - implementing the Local Search Algorithm with simple shifting move mmLSAm,
- OAmmTSAe - implementing the Tabu Search Algorithm with exchange move mmTSA,
- OAmmCA - implementing the Crossover Algorithm mmCA,
- OAmmPRA - implementing the Path Relinking Algorithm mmPRA.

Similarly to the optimization algorithms and optimization agents the cooperation strategy has been evolving during the research development. The modified implementations of algorithms for the MRCPSP arose from the A-Team approaches proposed for the RCPSP/max and MRCPSP/max developed from 2008 to 2011. These algorithms are described in [73] for the RCPSP/max and MRCPSP/max and in [72, 74–76] for the RCPSP and MRCPSP. The details of the basic static cooperation strategy with blocking can be found in Sect. 5.4.3 and [72].

5.4.3 A-Team with Dynamic Cooperation Strategy with Reinforcement Learning

Both the above described variants of A-Team solving instances of the MRCPSP use the static cooperation strategies. As in case of the RCPSP, there is a space for the performance improvements through using the dynamic strategies. Idea of applying the dynamic strategy with reinforcement learning to the MRCPSP originates from [72] where DCRL Strategy was proposed and experimentally validated. The DCRL Strategy approach has been described in Sect. 4.4.4 for the RCPSP. In case of the MRCPSP the strategy has been extended with the fourth rule. Hence, the following rules are considered:

- RL1 - controls replacement of one individual from the population by another, randomly generated one.
- RL2 - controls the method of selecting an individual for replacement.
- RL3 - controls the process of clustering individuals in the population according to certain features, and next selecting a cluster.
- RL4 - controls the procedure of selecting individuals to be forwarded to optimization agents from the cluster chosen by RL3.

The first three rules, as well as probabilities and weights used, are described in Sect. 4.4.4. The RL4 rule is complementary to the RL3. For each individual from the population the matrix of weights is remembered. It reflects results of the subsequent improvement attempts. In the considered case the matrix of four weights is used, one for each OptiAgent. The initial weights are identical. After receiving an "improved" individual from the OptiAgent the respective weight is decreased when the value of solution is worse and increased when this value is better or equal to the one before the attempted improvement took place. The probability of selecting an individual for improvement is calculated as

$$P_S^{Alg} = \frac{w_S^{Alg}}{\sum_{i \in G_{RL3}} w_i^{Alg}},$$

where G_{RL3} is the group of individuals indicated by RL3.

The rules RL1 and RL2 are used consecutively just before adding the new solution to the population after receiving it by the SolutionManager from the OptiAgent. The RL3 and RL4 rules are used just before reading the solution from the population. The RL3 and RL4 act asynchronously to the RL1 and RL2.

The proposed four rules are integrated with the static blocking strategy extending it to the dynamic one (see Sect. 4.4.4). Furthermore, the following combinations of the control elements are considered:

- RL34 - RL3 and RL4.
- RL123 - RL1 and RL2 and RL3.
- RL1234 - RL1 and RL2 and RL3 and RL4.

Each single rule as well as each of their combinations may be used in a separate dynamic cooperation strategy. The details of this approach can be found in [72].

In this book the RL123 and RL1234 strategy variants are considered, called DCRL Strategy and DCRL4 Strategy, respectively.

Computational experiments show that the proposed approaches are effective in solving instances of the MRCPSP. The best dynamic strategy with reinforcement learning uses all four rules (DCRL4 Strategy), but the differences as compared with other strategies are not significant. In comparison with DCRL Strategy the Mean RE and Max RE are slightly lower or equal in case of the data sets j16 and j20. However, for the DCRL4 Strategy approach the computational times are shorter. The results obtained by using these strategies are presented in Tables 5.4 and 5.5.

Table 5.4 Experiment results for the A-Team with DCRL Strategy solving the MRCPSP

Number of activities	Mean RE from opt*/bk solution [%]	Max RE from opt*/bk solution [%]	Mean RE from CPLB [%]	Max RE from CPLB [%]	Mean CT [s]	Mean TCT [s]
10	0.34	8.71	30.17	314.29	1.21	11.53
12	0.46	23.53	25.28	206.25	1.23	18.55
14	0.58	18.75	22.41	193.33	3.17	22.42
16	0.76	11.11	19.05	164.29	3.38	23.67
18	0.95	18.21	18.69	153.33	3.87	26.78
20	1.58	20.59	18.11	161.54	4.11	30.32

Table 5.5 Experiment results for the A-Team with DCRL4 Strategy solving the MRCPSP

Number of activities	Mean RE from opt*/bk solution [%]	Max RE from opt*/bk solution [%]	Mean RE from CPLB [%]	Max RE from CPLB [%]	Mean CT [s]	Mean TCT [s]
10	0.31	8.63	30.14	312.45	1.71	13.75
12	0.43	19.47	25.25	202.98	1.42	18.56
14	0.54	17.99	22.37	192.13	3.19	22.42
16	0.76	10.53	19.05	158.97	3.39	22.27
18	0.94	18.03	18.68	150.02	3.85	26.04
20	1.58	19.80	18.11	154.73	4.17	30.32

5.4.4 A-Team with Dynamic Cooperation Strategy Based on Population Learning

Further studies of the JABAT based A-Team implementations for solving the MRCPSP instances have involved cooperation strategies mainly. In this section the approach based on the population learning algorithm (PLA) is described. The DCPL Strategy proposed for RCPSP in Sect. 4.4.5 can be adopted to the multiple mode case. The differences are in using the JABAT classes and in the optimization algorithms implementation. Moreover, two modified versions of mmLSA and mmTSA algorithms are used: mmLSAe where the exchange move is used, and mmTSAm where the simple shifting move is used, similarly as in case of the RCPSP. The algorithms are implemented as optimization agents and additional OAmmLSAe and OAmmTSAe agents are added. Under the above described assumptions the following optimization agents are used:

- OAmmLSAm - implementing the Local Search Algorithm mmLSAm,
- OAmmLSAe - implementing the Local Search Algorithm mmLSAe,
- OAmmTSAm - implementing the Tabu Search Algorithm mmTSAm,
- OAmmTSAe - implementing the Tabu Search Algorithm mmTSAe,
- OAmmCA - implementing the Crossover Algorithm mmCA,
- OAmmPRA - implementing the Path Relinking Algorithm mmPRA.

To validate the approach three PLA based strategies have been implemented and tested, including 1, 2 or 3 learning stages. In each stage a different set of parameters and optimisation agents is used. In the PLA more advanced stages are entered by a diminishing number of individuals from the initial population. The best one, as in case of the RCPSP, occurs strategy using all three learning stages called DCPL Strategy. The sets of settings for this strategy, the same as in case of the RCPSP, are presented in Table 5.7. The values in brackets for optimization agents denote numbers of iterations of the respective optimization algorithms according to their descriptions in Sect. 5.4.2. For the LSA it is the total number of iterations, and for the TSA the maximum number of iterations in which no improvement is found. The *nSGS* denotes the number of SGS procedure calls performed by the algorithm. The avgDiv(P) denotes average diversity in the population.

Computational experiment shows that the proposed DCPL Strategy approach is efficient. The results are presented in Table 5.6. With respect to the mean relative error they are better than the results obtained by DCRL4 Strategy by an average of 23.41%, and DCRL Strategy by an average of 26.64%. The best improvement has been obtained for instances from j16 (44.7%) and j12 (32.6%) data sets. The least for instances from j14 (8.6%) data set.

Table 5.6 Parameter values for three learning stages used in DCPL Strategy

Stage 1	
Initial weights	$w_{mgr} = 50$, $w_{mgrc} = 50$, $w_{mgb} = 0$, $w_{mgbc} = 0$ $w_{mrr} = 100$, $w_{mrw} = 0$, $w_{mrt} = 0$
Optimization agents	OAmmCA, OAmmPRA
Selection criteria	$\alpha = 0.5$
Stopping criteria	avgDiv(P) < 0.1 and $nSGS$ > 2000
Stage 2	
Initial weights	$w_{mgr} = 25$, $w_{mgrc} = 25$, $w_{mgb} = 25$, $w_{mgbc} = 25$ $w_{mrr} = 34$, $w_{mrw} = 33$, $w_{mrt} = 33$
Optimization agents	OAmmLSAm($5n$), OAmmLSAe($5n$)
Selection criteria	$\alpha = 0.5$
Stopping criteria	avgDiv(P) < 0.5 and $nSGS$ > 1500
Stage 3	
Initial weights	$w_{mgr} = 0$, $w_{mgrc} = 0$, $w_{mgb} = 50$, $w_{mgbc} = 50$ $w_{mrr} = 0$, $w_{mrw} = 20$, $w_{mrt} = 80$
Optimization agents	OAmmTSAm($2n$), OAmmTSAe($2n$)
Selection criteria	$\alpha = 0.8$
Stopping criteria	avgDiv(P) < 0.05 and $nSGS$ > 1500

Table 5.7 Experiment results for the A-Team with DCPL Strategy solving the MRCPSP

Number of activities	Mean RE from opt*/bk solution [%]	Max RE from opt*/bk solution [%]	Mean RE from CPLB [%]	Max RE from CPLB [%]	Mean CT [s]	Mean TCT [s]
10	0.25	8.7	30.05	314.29	1.28	13.05
12	0.31	23.53	25.12	206.25	1.29	20.32
14	0.53	18.75	22.36	193.33	3.31	22.69
16	0.42	10.00	18.63	164.29	3.53	24.64
18	0.71	12.90	18.35	153.33	4.28	26.15
20	1.23	11.02	17.62	161.54	4.46	31.47

5.4.5 *A-Team with Dynamic Cooperation Strategy Based on Integration*

Good performance of both so far discussed dynamic cooperation strategies: DCRL Strategy and DCPL Strategy, stood behind an attempt to integrate them expecting a synergetic effect. The idea was proposed in [77] where the Dynamic Cooperative Interaction Strategy (DCI Strategy) for the RCPSP was described. To standardize terminology the strategy is called the Dynamic Cooperation Strategy based on the

Table 5.8 Experiment results for the A-Team with DCI Strategy solving the MRCPSP

Number of activities	Mean RE from opt*/bk solution [%]	Max RE from opt*/bk solution [%]	Mean RE from CPLB [%]	Max RE from CPLB [%]	Mean CT [s]	Mean TCT [s]
10	0.23	8.01	30.00	300.43	1.29	12.96
12	0.29	23.53	25.09	206.25	1.33	21.08
14	0.48	12.50	22.31	193.33	3.25	21.73
16	0.37	7.59	18.56	164.29	3.47	21.20
18	0.67	12.91	18.30	153.33	4.10	25.37
20	1.12	9.96	17.47	161.54	4.39	31.27

Integration of the best rules from previous approaches, notation remains the same - DCI Strategy.

In Sect. 4.4.6 the DCI Strategy for solving RCPSP has been described where the best ideas from RL and PL based strategies are combined together. A similar approach with the JABAT structure implementation for the multi-mode RCPSP can be used to solve the MRCPSP instances.

Like in the population learning based strategy six optimization agents, representing six optimization algorithms are used: mmLSAm, mmLSAe, mmTSAm and mmTSAe, mmCA, mmPRA.

The general schema of the DCI Strategy, as well as methods of generating a new individual and methods of replacing an individual from the population by a new one are described in Sect. 4.6.6 and [77]. They remain the same for the single and multi-mode problem.

The proposed approach has been validated experimentally, and the results are shown in Table 5.8. It can be observed that the A-Team using DCI Strategy approach is effective in solving MRCPSP. It also appears that the DCI Strategy approach is more efficient than the DCRL, DCRL4 and DCPL Strategy approaches. The results for DCI Strategy are improved by an average of 8.4% in comparison with DCPL Strategy and of 29.8% in comparison with DCRL4 Strategy. The best improvement has been obtained for instances from the test set j16 (11.9, 51.3%) with 16 activities.

5.4.6 Concluding Remarks

Computational experiments results show that proposed A-Team architectures are effective for solving the MRCPSP. The Mean Relative Errors (Mean RE) from optimal results presented in this section are summarized in Table 5.9. The graphical representation of these results are shown in Fig. 5.9. The most effective approaches are A-Teams with PLA-Strategy and DCI Strategy. In both these approaches six optimization algorithms and six optimization agents representing these algorithms have been used: mmCA, mmPRA, mmLSAm, mmLSAe, mmTSAm and mmTSAe.

To check whether there is a significant difference between the considered A-Teams (i.e. cooperation strategies) the non-parametric Friedman test has been used. It makes possible to answer the question weather particular working strategies are equally effective independently of the kind of problem being solved. Additionally, the mean ranks calculated for each A-Team have been inspected to roughly evaluate the effectiveness of the A-Teams.

The test is based on ranks assigned to each A-Team participating in the experiment, i.e. A-Team with SC1, SC2, DCRL, DCRL4, DCPL, and DCI Strategy. To assign ranks, the 6 point scale was used, with 6 points for the best and 1 point for the worst result found by the A-Team for a particular problem instance. When the results are identical, the same amount of points equaled the mean of consecutive ranks is assigned to each such result (i.e. there are ties among ranks).

The test aimed at deciding among the following hypotheses:

Table 5.9 Mean RE from the optimal solution for JABAT-based A-Team approaches to solve MRCPSP discussed in this section

Number of activities	SC1 Strategy [%]	SC2 Strategy [%]	DCRL Strategy [%]	DCRL4 Strategy [%]	DCPL Strategy [%]	DCI Strategy [%]
10	0.75	0.69	0.34	0.31	0.25	0.23
12	0.76	0.70	0.46	0.43	0.31	0.29
14	0.80	0.76	0.58	0.54	0.53	0.48
16	0.83	0.79	0.76	0.76	0.42	0.37
18	0.98	0.91	0.95	0.94	0.71	0.67
20	1.85	1.81	1.58	1.58	1.23	1.12

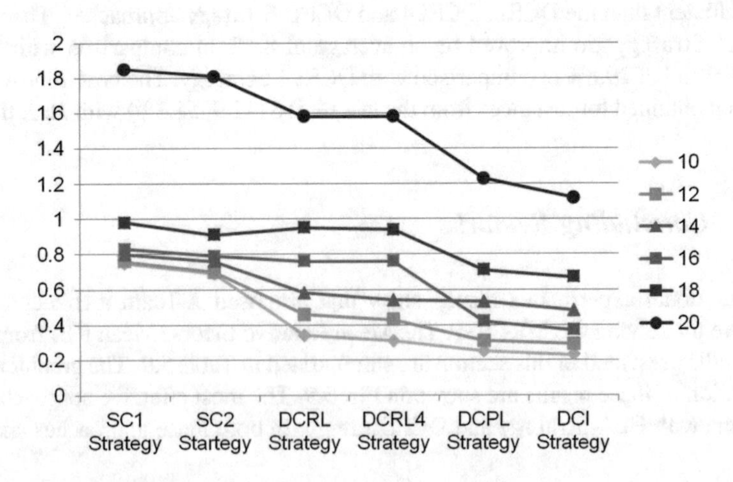

Fig. 5.9 Graphical representation of the results from Table 5.9

Table 5.10 Values of the χ^2 statistics for the MRCPSP

Number of activities	χ^2
10	115.13
12	98.33
14	81.27
16	1746.80
18	47.67
20	1158.42

H_0 zero hypothesis: A-Teams (cooperation strategies) are statistically equally effective i.e. obtain similar results with nonsignificant differences,
H_1 alternative hypothesis: not all A-Teams (cooperation strategies) are equally effective.

The test has been performed on 6 cooperation strategies and 6 data sets of the considered multi-mode RCPSP from PSPLIB: j10, j12, j14, j16, j18, and j20, including 536, 547, 551, 550, 552, and 554 instances, respectively. The analysis has been carried out at the significance level of 0.05 and 5 degrees of freedom. The respective values of the χ^2 statistics are shown in Table 5.10.

The critical value of χ^2 distribution for the assumed values equals 11.07. Since the obtained values of the statistics χ^2, in all cases, are greater than the critical one, hypothesis H_0 is rejected. Thus, the obtained result proves hypothesis H_1 which claims that not all A-Teams are equally effective in solving the MRCPSP instances.

The mean values of ranks, calculated for the considered A-Team, might suggest some ranking with respect to their efficiency. These are shown in the Fig. 5.10.

As it could be seen from the Fig. 5.10, in all cases two A-Teams can be considered as slightly superior. They have similar mean ranks, although the A-Team with DCI Strategy is always the better one. Additionally, in case of the instances including 10, 12, and 14 activities four A-Teams with dynamic cooperation strategies (DCRL, DCPL DCRL4 and DCI) are clearly visible as superior to the A-Teams with static strategies. In case of the data set including instances with 16 activities the mean ranks for the A-Teams with DCPL Strategy and DCI Strategy are noticeably higher.

Computational experiments allow to state that the proposed approaches to solving MRCPSP using A-Team implementation can be considered as useful and competitive tools for solving instances of the MRCPSP. The obtained results are comparable with the best results known from the literature. Such comparison is presented in Table 5.11. The presented comparison includes results from the other multi-agent system using reinforcement learning, denoted as MAS with RL, which has been proposed recently by Wauters et al. [61] (see Sect. 5.3). It gives slightly better results. In MAS with RL approach the similar methods based on reinforcement learning have been used, but the structure of the system is different. Each agent represents an activity and uses learning automata.

Other approaches from the literature includes: distribution algorithm, genetic algorithm, scatter search and hybrid methods. All of them uses the same test sets

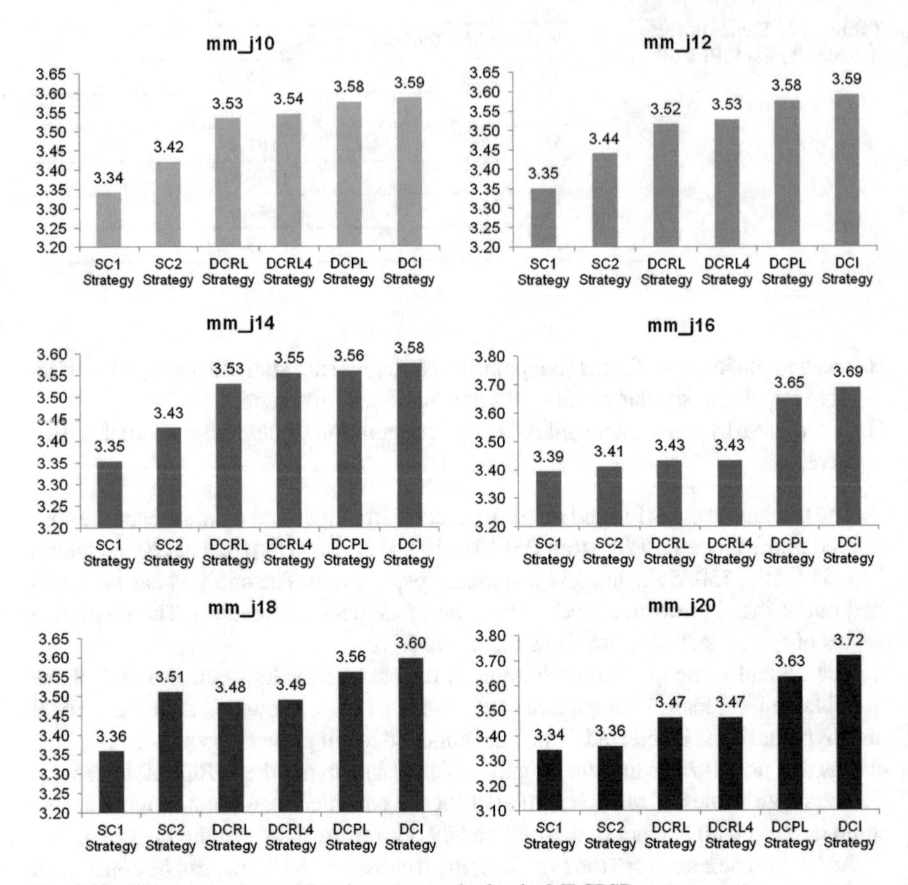

Fig. 5.10 The mean values of Friedman test ranks for the MRCPSP

from PSPLIB and all of them produce quite similar results. The differences between the Mean RE reported in the literature for the presented algorithms are small: from 0.22% for the test set j10 to 1.32% for the test set j20. The detailed and honest comparison is difficult because of the different systems, platforms and machines (processors) used. In respect to the practical applications the differences for the larger projects are more significant, so the proposed approaches for projects with higher number of activities should be tested and compared. It is worth mentioning that in agent based approaches computations are performed by many processors (nodes) working in parallel. Additionally, computation times include times used by agents to prepare, send and receive messages.

Table 5.11 Literature reported results [32, 34, 35, 61]

Set	Algorithm	Authors	Mean RE [%]	Mean CT [s]	CPU
mm10	Distribution algorithm	Wang and Fang	0.01	1	2.2 GHz
	Genetic algorithm	Van Peteghem and Vanhoucke	0.01	0.12	2.8 GHz
	Hybrid genetic algorithm	Lova et al.	0.04	0.1	3 GHz
	MAS with RL	Wauters et al.	0.05	0.8	2.8 GHz
	A-Team with DCI Strategy		*0.23*	*1.29*	*1.4* GHz
mm12	Distribution algorithm	Wang and Fang	0.02	1.8	2.2 GHz
	Genetic algorithm	Van Peteghem and Vanhoucke	0.09	–	–
	MAS with RL	Wauters et al.	0.08	1.0	2.8 GHz
	Hybrid genetic algorithm	Lova et al.	0.17	–	–
	A-Team with DCI Strategy		*0.29* GHz	*1.33*	*1.4*
	Hybrid scatter search	Ranjbar et al.	0.65	10	3 GHz
mm14	Distribution algorithm	Wang and Fang	0.03	1	2.2 GHz
	Genetic algorithm	Van Peteghem and Vanhoucke	0.22	0.14	2.8 GHz
	MAS with RL	Wauters et al.	0.23	1.4	2.8 GHz
	Hybrid genetic algorithm	Lova et al.	0.32	0.11	3 GHz
	A-Team with DCI Strategy		*0.48*	*3.25*	*1.4* GHz
	Hybrid scatter search	Ranjbar et al.	0.89	10	3 GHz
mm16	Distribution algorithm	Wang and Fang	0.17	1	2.2 GHz
	Genetic algorithm	Van Peteghem and Vanhoucke	0.32	0.15	2.8 GHz
	MAS with RL	Wauters et al.	0.30	1.7	2.8 GHz
	A-Team with DCI Strategy		*0.37*	*3.47*	*1.4* GHz
	Hybrid genetic algorithm	Lova et al.	0.44	0.12	3 GHz
	Hybrid scatter search	Ranjbar et al.	0.95	10	3 GHz
mm18	Distribution algorithm	Wang and Fang	0.19	1	2.2 GHz
	Genetic algorithm	Van Peteghem and Vanhoucke	0.42	0.16	2.8 GHz
	MAS with RL	Wauters et al.	0.53	1.9	2.8 GHz
	Hybrid genetic algorithm	Lova et al.	0.63	0.13	3 GHz
	A-Team with DCI Strategy		*0.67*	*4.10*	*1.4* GHz
	Hybrid scatter search	Ranjbar et al.	1.21	10	3 GHz
mm20	Distribution algorithm	Wang and Fang	0.32	1	2.2 GHz
	Genetic algorithm	Van Peteghem and Vanhoucke	0.57	0.17	2.8 GHz
	MAS with RL	Wauters et al.	0.70	2.1	2.8 GHz
	Hybrid genetic algorithm	Lova et al.	0.87	0.15	3 GHz
	A-Team with DCI Strategy		*1.12*	*4.39*	*1.4* GHz
	Hybrid scatter search	Ranjbar et al.	1.64	10	3 GHz

References

1. Talbot, F.B.: Resource-constrained project scheduling with time-resource trade-offs: the non preemptive case. Manag. Sci. **28**(10), 1197–1210 (1982)
2. Błażewicz, J., Lenstra, J., Rinnooy, A.: Scheduling subject to resource constraints: classification and complexity. Discret. Appl. Math. **5**, 11–24 (1983)
3. Kölisch, R.: Project scheduling under resource constraints — efficient heuristics for several problem classes. Ph.D. thesis, Physica, Heidelberg (1995)
4. Brucker, P., Drexl, A., Möhring, R., Neumann, K., Pesch, E.: Resource-constrained project scheduling: notation, classification, models, and methods. Eur. J. Oper. Res. **112**, 3–41 (1999)
5. Demeulemeester E., Herroelen W.: Project Scheduling: A Research Handbook. Kluwer Academic Publishers (2002)
6. Boctor, F.: Heuristics for scheduling projects with resource restrictions and several resource-duration modes. Int. J. Prod. Res. **31**, 2547–2558 (1993)
7. Boctor, F.: A new and efficient heuristic for scheduling projects with resource restrictions and multiple execution modes. Eur. J. Oper. Res. **90**, 349–361 (1996)
8. Sprecher, A.: Resource-Constrained Project Scheduling: Exact Methods for the Multi-Mode Case. Springer (1994)
9. Hartmann, S., Drexl, A.: Project scheduling with multiple modes: a comparison of exact algorithms. Networks **32**, 283–297 (1998)
10. Herroelen, W.S., De Reyck, B., Demeulemeester, E.L.: Resource-constrained project scheduling: a survey of recent developments. Comput. Oper. Res. **25**(4), 279–302 (1998)
11. Kölisch, R., Padman, R.: An integrated survey of deterministic project scheduling. OMEGA Int. J. Manag. Sci. **29**(3), 249–272 (2001)
12. Hartmann, S., Briskorn, D.: A survey of variants and extensions of the resource-constrained project scheduling problem. Eur. J. Oper. Res. **207**, 1–14 (2010)
13. Węglarz, J., Józefowska, J., Mika, M., Waligóra, G.: Project scheduling with finite or infinite number of activity processing modes – a survey. Eur. J. Oper. Res. **208**, 177–205 (2011)
14. Peteghem, V.V., Vanhoucke, M.: An experimental investigation of metaheuristics for the multi-mode resource-constrained project scheduling problem on new dataset instances. Eur. J. Oper. Res. **235**(1), 62–72 (2014)
15. Mika, M., Waligóra, G., Węglarz, J.: Overview and state of the art. In: Handbook on Project Management and Scheduling, vol. 1, pp. 445–490. Springer International Publishing (2015)
16. Słowiński, R.: Two approaches to problems of resource allocation among project activities – a comparative study. J. Oper. Res. Soc. **8**, 711–723 (1980)
17. Patterson, J., Słowiński, R., Talbot, F., Węglarz, J.: An algorithm for a general class of precedence and resource constrained scheduling problem. In: Advances in Project Scheduling, pp. 3–28. Elsevier, Amsterdam (1989)
18. Speranza, M., Vercellis, C.: Hierarchical models for multi-project planning and scheduling. Eur. J. Oper. Res. **64**, 312–325 (1993)
19. Hartmann, S., Sprecher, A.: A note on hierarchical models for multi-project planning and scheduling. Eur. J. Oper. Res. **94**, 377–383 (1996)
20. Sprecher, A., Hartmann, S., Drexl, A.: An exact algorithm for the project scheduling with multiple modes. OR Spectr. **19**, 195–203 (1997)
21. Sprecher, A., Drexl, A.: Solving multi-mode resource-constrained project scheduling problems by a simple, general and powerful sequencing algorithm. Eur. J. Oper. Res. **107**, 431–450 (1998)
22. Zhu, G., Bard, J., Tu, G.: A branch-and-cut procedure for the multimode resource-constrained project-scheduling problem. J. Comput. **18**(3), 377–390 (2006)
23. Drexl, A., Grünewald, J.: Nonpreemptive multi-mode resource-constrained project scheduling. IIE Trans. **25**, 74–81 (1993)
24. Özdamar, L., Ulusoy, G.: A local constraint based analysis approach to project scheduling under general resource constraints. Eur. J. Oper. Res. **79**, 287–298 (1994)

25. Özdamar, L., Ulusoy, G.: A survey on the resource-constrained project scheduling problem. IIE Trans. **27**(5), 574–586 (1995)
26. Kölisch, R., Drexl, A.: Local search for nonpreemptive multi-mode resource-constrained project scheduling. IIE Trans. **29**, 987–999 (1997)
27. Mori, M., Tseng, C.: A genetic algorithm for the multi-mode resource constrained project scheduling problem. Eur. J. Oper. Res. **100**, 134–141 (1997)
28. Özdamar, L.: A genetic algorithm approach to a general category project scheduling problem. IEEE Trans. Syst. Man Cybern. **29**(1), 44–59 (1999)
29. Hartmann, S.: Project scheduling with multiple modes: a genetic algorithm. Ann. Oper. Res. **102**, 111–135 (2001)
30. Alcaraz, J., Maroto, C., Ruiz, R.: Solving the multi-mode resource-constrained project scheduling problem with genetic algorithms. J. Oper. Res. Soc. **54**(6), 614–626 (2003)
31. Tseng, L.-Y., Chen, S.-C.: Two-phase genetic local search algorithm for the multimode resource-constrained project scheduling problem. IEEE Trans. Evol. Comput. **13**, 848–857 (2009)
32. Peteghem, V.V., Vanhoucke, M.: A genetic algorithm for the preemptive and non-preemptive multi-mode resource-constrained project scheduling problem. Eur. J. Oper. Res. **201**, 409–418 (2010)
33. Coelho, J., Vanhoucke, M.: Multi-mode resource-constrained project scheduling using RCPSP and SAT solvers. Eur. J. Oper. Res. **213**, 73–82 (2011)
34. Lova, A., Tormos, P., Cervantes, M., Barber, F.: An efficient hybrid genetic algorithm for scheduling projects with resource constraints and multiple execution modes. Int. J. Prod. Econ. **117**, 302–316 (2009)
35. Ranjbar, M., De Reyck, B., Kianfar, F.: A hybrid scatter-search for the discrete time/resource trade-off problem in project scheduling. Eur. J. Oper. Res. **193**, 35–48 (2009)
36. Peteghem, V.V., Vanhoucke, M.: Using resource scarceness characteristics to solve the multi-mode resource-constrained project scheduling problem. J. Heuristics **17**(6), 705–728 (2011)
37. Słowiński, R., Soniewicki, B., Węglarz, J.: DSS for multiobjective project scheduling. Eur. J. Oper. Res. **79**, 220–229 (1994)
38. Józefowska, J., Mika, M., Różycki, R., Waligóra, G., Węglarz, J.: Simulated annealing for multi-mode resource-constrained project scheduling. Ann. Oper. Res. **102**, 137–155 (2001)
39. Bouleimen, K., Lecocq, H.: A new efficient simulated annealing algorithm for the resource-constrained project scheduling problem and its multiple modes version. Eur. J. Oper. Res. **149**, 268–281 (2003)
40. Zhang, H., Tam, C.M., Li, H.: Multi-mode project scheduling based on particle swarm optimization. Comput. Aided Civ. Infrastruct. Eng. **21**, 93–103 (2006)
41. Jarboui, B., Damak, N., Siarry, P., Rebai, A.: A combinatorial particle swarm optimization for solving multi-mode resource-constrained project scheduling problems. Appl. Math. Comput. **195**, 299–308 (2008)
42. Chiang, C., Huang, Y., Wang, W.: Ant colony optimization with parameter adaptation for multi-mode resource-constrained project scheduling. J. Intell. Fuzzy Syst. **29**, 345–358 (2008)
43. Damak, N., Jarboui, B., Siarry, P., Loukil, T.: Differential evolution for solving multi-mode resource-constrained project scheduling problems. Comput. Oper. Res. **36**, 2653–2659 (2009)
44. Tchao, C., Martins, S.L.: Hybrid heuristics for multi-mode resource-constrained project scheduling. In: Maniezzo, V., Battiti, R., Watson, J.P. (eds.) Learning and Intelligent Optimization (LION 2007). Lecture Notes in Computer Science, vol. 5313, pp. 234–242 (2008)
45. Wang, L., Fang, C.: An effective shuffled frog-leaping algorithm for multi-mode resource-constrained project scheduling problem. Special Issue on Interpretable Fuzzy Systems. Inf. Sci. **181**(20), 4804–4822 (2011)
46. Kyriakidis, T.S., Kopanos, G.M., Georgiadis, M.C.: MILP formulations for single- and multi-mode resource-constrained project scheduling problems. Comput. Chem. Eng. **36**, 369–385 (2012)
47. Wang, L., Fang, C.: An effective estimation of distribution algorithm for the multi-mode resource-constrained project scheduling problem. Comput. Oper. Res. **39**, 449–460 (2012)

48. Li, H., Zhang, H.: Ant colony optimization-based multi-mode scheduling under renewable and nonrenewable resource constraints. Autom. Constr. **35**, 431–438 (2013)
49. Sebt, M.H., Afshar, M.R., Alipouri, Y.: An efficient genetic algorithm for solving the multi-mode resource-constrained project scheduling problem based on random key representation. Int. J. Supply Oper. Manag. **2**(3), 905–924 (2015)
50. Geiger, M.J.: A multi-threaded local search algorithm and computer implementation for the multi-mode, resource-constrained multi-project scheduling problem. Eur. J. Oper. Res. **256**(3), 729–741 (2017)
51. MMLIB - Multi-Mode project scheduling problem LIBrary. http://www.projectmanagement.ugent.be/research/data/RanGen
52. PSPLIB - Project Scheduling Problem LIBrary. http://www.om-db.wi.tum.de/psplib
53. Kölisch, R., Sprecher, A., Drexl, A.: Characterization and generation of a general class of resource-constrained project scheduling problems. Manag. Sci. **41**, 1693–1703 (1995)
54. Geiger, M.J.: MMLIB checker — a checker software for multi-mode resource-constrained project scheduling problem (MRCPSP) solution files. Research Report RR-15-03-01. Helmut-Schmidt-University/University of the Federal Armed Forces Hamburg, Logistics Management Department, Hamburg, Germany (2015)
55. Project data for integrated project management and control. http://www.projectmanagement.ugent.be/?q=research/data
56. Demeulemeester, E., Vanhoucke, M., Herroelen, W.: RanGen: a random network generator for activity-on-the-node networks. J. Sched. **6**(1), 17–38 (2003)
57. Knotts, G., Dror, M., Hartman, B.C.: Agent-based project scheduling. IIE Trans. **32**(5), 387–401 (2000)
58. Knotts, G., Dror, M.: Agent-based project scheduling: computational study of large problems. IIE Trans. **35**, 143–159 (2003)
59. Maroto, C., Tormos, P.: Project management: an evaluation of software quality. Int. Trans. Oper. Res. **1**, 209–221 (1994)
60. Wauters, T., Verbeeck, K., Berghe, G.V., De Causmaecker, P.: A multi-agent learning approach for the multi-mode resource-constrained project scheduling problem. In: Decker, S., Sierra, C. (eds.) Proceedings of 8th International Conference on Autonomous Agents and Multiagent Systems (AAMAS 2009), pp. 1–8. International Foundation for Autonomous Agents and Multiagent Systems. www.ifaamas.org
61. Wauters, T., Verbeeck, K., Berghe, G.V., De Causmaecker, P.: Learning agents for the multi-mode project scheduling problem. J. Oper. Res. Soc. **62**, 281–290 (2011)
62. Mirzaei, O., Akbarzadeh, T.R.M.: A novel learning algorithm based on a multi-agent structure for solving multi-mode resource-constrained project scheduling problem. J. Convergence **4**(1), 47–52 (2013)
63. Wenzler, F., Günthner, W.A.: A learning agent for a multi-agent system for project scheduling in construction. In: Claus, T., Herrmann, F., Manitz, M., Rose, O. (eds.) Proceedings of the 30th Conference on Modelling and Simulation, pp. 11–17 (2016)
64. Jędrzejowicz, P., Ratajczak-Ropel, E.: Agent-based approach to solving the resource constrained project scheduling problem. Lect. Notes Comput. Sci. **4431**, 480–487 (2007)
65. Jędrzejowicz, P., Ratajczak-Ropel, E.: New generation A-Team for solving the resource constrained project scheduling. In: Proceedings of the Eleventh International Workshop on Project Management and Scheduling, pp. 156–159. Istanbul (2008)
66. Kölisch, R., Sprecher, A.: PSPLIB–A project scheduling problem library. Eur. J. Oper. Res. **96**, 205–216 (1996)
67. Glover, F., Laguna, M.: Tabu Search. Kluwer Academic Publishers (1997)
68. Glover, F.: Tabu search - Part I. ORSA J. Comput. **1**, 190–206 (1989)
69. Glover, F.: Tabu search - Part II. ORSA J. Comput. **2**, 4–32 (1989)
70. Glover, F.: Tabu search and adaptive memory programing: advances, applications and challenges. In: Barr, R.S., Helgason, R.V., Kennington, J.L. (eds.) Interfaces in Computer Scinece and Operations Research, pp. 1–75. Kluwer (1996)

71. Glover, F., Laguna, M., Marti, R.: Fundamentals of scatter search and path relinking. Control Cybern. **39**, 653–684 (2000)
72. Jędrzejowicz, P., Ratajczak-Ropel, E.: Reinforcement learning strategy for solving the MRCPSP by a team of agents. In: Neves-Silva, R., Jain, L.C., Howlett, R.J. (eds.) Intelligent Decision Technologies, Proceedings of the 7th KES International Conference on Intelligent Decision Technologies (KES-IDT 2015), pp. 537–548. Springer International Publishing, Switzerland (2015)
73. Jędrzejowicz, P., Ratajczak-Ropel, E.: Double-action agents solving the MRCPSP/Max problem. In: Jędrzejowicz, P., et al. (eds.) Computational Collective Intelligence. Technologies and Applications. Lecture Notes in Artificial Intelligence, vol. 6923, pp. 311–321 (2011)
74. Jędrzejowicz, P., Ratajczak-Ropel, E.: Team of A-Teams for solving the resource-constrained project scheduling problem. In: Grana, M., Toro, C., Posada, J., Howlett, R., Lakhmi, C.J. (eds.) Advances in Knowledge Based and Intelligent Information and Engineering Systems. Frontiers in Artificial Intelligence and Applications, vol. 243, pp. 1201–1210 (2012)
75. Jędrzejowicz, P., Ratajczak-Ropel, E.: Reinforcement learning strategy for solving the resource-constrained project scheduling problem by a team of A-Teams. In: Nguyen, N.T., Attachoo, B., Trawiński, B., Somboonviwat, K. (eds.) Intelligent Information and Database Systems. Lecture Notes in Artificial Intelligence, vol. 8398, pp. 197–206 (2014)
76. Jędrzejowicz, P., Ratajczak-Ropel, E.: Reinforcement learning strategies for A-Team solving the resource-constrained project scheduling problem. Neurocomputing **146**, 301–307 (2014)
77. Jędrzejowicz, P., Ratajczak-Ropel, E.: Dynamic cooperative interaction strategy for solving RCPSP by a team of agents. In: Nguyen, N.T., Manolopoulos, Y., Iliadis, L., Trawiński, B. (eds.) Computational Collective Intelligence. Lecture Notes in Artificial Intelligence, vol. 9875, pp. 454–463 (2016)

Chapter 6
Conclusions

In this part of the book the agent based approaches to single-mode and multi-mode resource-constrained project scheduling problem have been considered. A few such approaches i.e. Multi-Agent Systems for the RCPSP or MRCPSP have been proposed in the literature. There are based on different agent based architectures and assumptions about agents' roles in the system. All of them are promising and are able to generate good results, but implementation and configuration such systems requires a lot of effort. This conclusion also applies to the A-Team based multi-agent systems which have been mainly considered in the described research. In contrast to the others, in the A-Team based MAS agents represent the system entities like: optimisation algorithms, populations, managers controlling the process of solving, error monitoring, etc. A problem instance is solving as a whole by optimization agents representing optimization algorithms.

The research presented in this part of the book and in particular computational experiments carried-out to validate the proposed A-Team based approach allow for some general observations and conclusions. First of all, it should be noted that the proposed MAS based on the A-Team paradigm should be considered as a powerful tool for solving the single-mode and multi-mode resource-constrained project scheduling problems. Unfortunately, this tool is also difficult to configure and testing because of its complexity. Selection of optimization algorithms used, design of optimization agents, as well as selection and settings of the cooperation strategies influence, to a substantial extend, quality of the system performance. Selecting optimization agents requires maintaining reasonable balance between computation times required by the different optimization procedure. If computation times required by optimization agents to perform their tasks differ too much, the system is prone to problems with respect to coordination and communication. An agent generating solutions with a substantial speed while others are engaged in search for an improved solutions can cause unexpected disturbance. Similarly, an agent working too slowly can cause delays and weaker overall performance through reducing

E. Ratajczak-Ropel and A. Skakovski, *Population-Based Approaches to the Resource-Constrained and Discrete-Continuous Scheduling*,
Studies in Systems, Decision and Control 108, DOI 10.1007/978-3-319-62893-6_6

population diversity. It has been also observed that the expected synergetic effect while combining more effective and complex algorithms with simpler ones does not, usually, meet expectations. Reasonable solution is to select optimization procedures and to design optimization agents in such a way that their computation time performance is similar.

It seems also wise to consider blocking weak individuals more often or to a grater extend than better ones. A repeated attempt to improve such weak individuals may prevent improving the better ones. Future research could lead to solutions with different blocking strategies applied to different individuals.

Development of agent technologies could make the idea of using the A-Team paradigm to solving computationally hard combinatorial optimization problems even more attractive. It has been demonstrated that combining efforts of different agents using different procedures and being controlled by sophisticated strategies based on some machine learning tool can bring about synergetic effect boosting performance of the whole system. It is also worth considering to take advantage of cloud computing and further parallelization to assure easy and flexible access to A-Team services.

Part II
Population-Based Approaches to the Discrete-Continuous Scheduling

Aleksander Skakovski

Chapter 7
Introduction

The discrete-continuous scheduling problem (DCSP), discussed in the following part of this book, is a particular case of the resource-constrained project scheduling problem (RCPSP). For this reason we begin with a brief description of the RCPSP at the beginning of Chap. 8 and continue with a thorough discussion on the DCSP in the remainder. We proceed with a short review of practical applications of the DCSP in Sect. 8.2, followed by the notation and the task models description in Sects. 8.3 and 8.4 respectively. After that, we give formulation of the DCSP in Sect. 8.5, consider its variants in Sect. 8.6 and the general approach to solving the problem in Sect. 8.7. Further on, we continue with the main properties of optimal schedules in Sect. 8.8, which are very useful in the construction and analysis of scheduling algorithms, and can even lead to analytical results. In the DCSP, similarly to the more general RCPSP, each task requires certain amount of a single renewable resource to be performed. This resource is continuous, i.e. divisible continuously. The time and rate of processing the task depend on the unknown in advance amount of the continuous resource to be allocated to the task. Therefore, in order to solve the DCSP it is required to determine the sequence of tasks on the machines and the allocation of the continuous resource to the tasks such, that optimize given criterion. Because the amount of the continuous resource is not known in advance, the allocation of the continuous resource can be treated as a separate subproblem. In the DCSP, the processing rate of a task is a function of the amount of the continuous resource allocated to the task. Although, we consider the properties of both convex and concave processing rate functions in Sects. 8.8.1 and 8.8.2, however special attention is paid to the concave power processing rate functions in Sect. 8.8.3, which are most important from the practical point of view. In the study, we mainly discuss about solving the DCSP for the case of the makespan (the schedule length) C_{max} minimization, however, the cases of the DCSP for the maximum lateness L_{max} and the mean flow time $\bar{F}$ are considered as well in Sects. 8.9 and 8.10 respectively.

E. Ratajczak-Ropel and A. Skakovski, *Population-Based Approaches to the Resource-Constrained and Discrete-Continuous Scheduling*, Studies in Systems, Decision and Control 108, DOI 10.1007/978-3-319-62893-6_7

Although it is possible to solve some cases of the continuous resource allocation problem analytically, the DCSP itself is NP-hard in the general case. For this reason, a variety of approaches were developed to cope with the problem. In the state-of-the-art review provided in Chap. 9, one can find another formulation of the DCSP given in Sect. 9.1.1, the new approach to optimal continuous resource allocation in Sect. 9.1.2, the new properties of the discrete part of the DCSP in Sect. 9.1.3, and the description of the discretisation of the DCSP in Sect. 9.2. Since, the DCSP is computationally intractable in the general case, the optimal approach for solving the problem is computationally ineffective or even impossible in practice. Instead, heuristic and metaheuristic approaches are used to manage both sequencing the tasks on the machines and allocation of the continuous resource. A review of existing heuristic and metaheuristic algorithms developed for solving the DCSP is provided in Sects. 9.3 and 9.4. In addition, there is also a review of the research on the minimization of the continuous resource usage in Sect. 9.5, in particular, the attention was paid to the energy consumption minimization. In Sect. 9.6, we also consider a continuous resource sharing (CRSharing) problem, which is a special case of the DCSP dealing with the continuous resource assignment to a given sequence of tasks on machines. At the end of the state-of-the-art review, we provide in Sect. 9.7 a survey of the research on the island model, often exploited in evolutionary computation, and the research on preventing premature convergence in evolutionary and genetic algorithms in Sect. 9.8.

The culminations of the study are Chaps. 10 and 11. In Chap. 10, we present proposed earlier metaheuristic evolutionary algorithms for solving the DCSP with continuous resource discretisation. In Sect. 10.1, we presented the island-based evolutionary algorithm with homogeneous islands (IBEA), in Sect. 10.2—the population learning algorithm (PLA), which implements the analogy to a social learning and introduces heterogeneity into the island model. The idea of social learning framework combined with the heterogeneity of the island model was further developed in the cross-entropy-based PLA2, presented in Sect. 10.3, and the PLA3 enhanced by Differential Evolution and presented in Sect. 10.4. The idea of the Differential Evolution based on a homogeneous island model was implemented in the island-based differential evolution algorithm (IBDEA) presented in Sect. 10.5.

In Chap. 11, the efficiency and properties of the algorithms presented in Chap. 10 were investigated. In Sect. 11.1, the performance of these algorithms was compared using the Friedman test. In Sect. 11.2, we studied the relations between the structure of the PLA2 and its efficiency. In Sect. 11.3, we examined the properties and the performance of the DE search based on a single population and the DE search based on the island model. We investigated how the effectiveness of the models depends on such parameters as the size of a single population, and in the case of the island model, also the number of islands and the migration rate. In Sect. 11.4, we also investigated the extent to which the performance of a considered differential evolution algorithm (DEA) depends on such parameters as the population diversification rate, the size of the population, and the number of fitness function evaluations. In Sect. 11.4.2, we described a decloning procedure, which was used for cyclic diversification of the population, and in Sect. 11.4.7—a performance improvement policy, based on the experimentally determined properties of the DEA. In Chap. 12, we finalize our study with conclusions.

Chapter 8
Discrete-Continuous Scheduling Problem

8.1 General Resource-Constrained Scheduling Problem

The discrete-continuous scheduling problem (DCSP) is a particular case of the more general resource-constrained project scheduling problem (RCPSP). The detailed description, as well as a survey of variants and extensions of the RCPSP, the reader might find in [1].

The RCPSP deals with a project consisting of activities $\boldsymbol{J} = \{J_1, J_2, \ldots, J_n\}$. Once started, activity may not be interrupted, i.e., activities are nonpreemtable. Due to technological requirements, there might be precedence relations among the activities. These relations are determined by sets of immediate predecessors of the activities $Pred_i$ indicating that an activity J_i may not be started before all of its predecessors are completed. The precedence relations can be represented by an activity-on-node network which is assumed to be acyclic. Each activity requires certain amounts of scarce resources to be performed. The resources are called renewable because their full capacity is available in every period. There are K renewable resources labeled $k = 1, \ldots, K$. It is assumed that the availability R_k of each resource k is constant for each period over time. Activity J_i requires r_{ik} units of resource k in each period of its processing. Thus, each activity of the project is characterized by its duration, resource requests, and precedence relations among the activities. Usually, two additional "dummy" activities J_0 and J_{n+1} are added in order to represent the start and the completion points of the project respectively. The duration of J_0 and J_{n+1} equals 0 and they do not require any resource. All information is assumed to be deterministic and known in advance. The parameters are assumed to be nonnegative and integer valued. A schedule is an assignment of start times S_i to the activities J_i, $i = 0, 1, \ldots, n + 1$. The objective is to find a schedule which leads to the earliest possible end of the project, i.e. the minimal makespan, and satisfies all precedence and resource constraints. It has been proven in [2] that the RCPSP belongs to the class of the strongly NP-hard problems.

E. Ratajczak-Ropel and A. Skakovski, *Population-Based Approaches to the Resource-Constrained and Discrete-Continuous Scheduling*,
Studies in Systems, Decision and Control 108, DOI 10.1007/978-3-319-62893-6_8

In the RCPSP, described above, the considered resources were available in discrete quantities only. Węglarz et al. in [3] generalized the concept of renewable resources by allowing continuously divisible resources. A methodology for discrete-continuous scheduling was proposed in [4]. Before we formulate and discuss the DCSP in more details, we will give some examples of practical situations where the DCSP arises.

8.2 Practical Applications of the DCSP

There are practical situations in which additional resources can be allocated to tasks in certain amounts (unknown in advance) within given intervals. Such resources may be called continuously divisible or simply continuous. These may be, among others, the situations when tasks are assigned to parallel processors driven by a common (electric, hydraulic, pneumatic) power source, e.g. commonly supplied grinding or mixing machines, electrolytic tanks or refueling terminals. As another example, one can consider the forging process in steel plants [5]. Forgings are preheated by gas up to an appropriate temperature in forge furnaces. Gas flow intensity, limited for the whole battery of forge furnaces, is a continuous resource. As another example one can consider manpower or money being a continuous resource. Also, in computer systems, multiple processors may share a common primary memory. If it is a paged virtual memory system and the number of pages goes into hundreds, primary memory can be treated as a continuous resource, see [6]. On the other hand, in scalable (SPP) and massively parallel (MPP) systems with hundreds or even thousands of processors, processors themselves can be considered as the continuous resource and the role of the machines can be played e.g. by disk drives.

In the following subsection we introduce the notions and task models, which will be used for the description of the DCSP.

8.3 Notation

Below, the notation used for the description of the DCSP is given in as follows:

C_i	the completion time of task J_i (unknown in advance), $i = 1, 2, \ldots, n$,
d_i	the due date of task J_i, $i = 1, 2, \ldots, n$,
$f_i(\bullet)$	continuous non-decreasing function, $f_i(0) = 0$, $i = 1, 2, \ldots, n$,
$g_i(\bullet)$	continuous, non-negative, non-increasing function, $g_i(0) = \infty$, $i = 1, 2, \ldots, n$,

$\boldsymbol{J} = \{J_1, J_2, \ldots, J_n\}$	a set of n independent, nonpreemptable tasks,
l_i	processing mode l of task J_i, $l_i = 1, 2, \ldots, W_i$,
M_k	time intervals, defined by the completion times of consecutive tasks, $k = 1, 2, \ldots, p$, $p \le n$,
$\boldsymbol{P} = \{P_1, P_2, \ldots, P_m\}$	a set of m parallel and identical machines (the discrete resource),
r_i	the release date of task J_i, $i = 1, 2, \ldots, n$,
S^z	a feasible sequence of combinations Z_k associated with each feasible schedule, $k = 1, 2, \ldots, p$,
τ_i	the processing time of task J_i, $i = 1, 2, \ldots, n$,
$\tau_i^{l_i}$	processing time of task J_i in mode $l_i = 1, 2, \ldots, W_i$,
$\boldsymbol{\tau}_i = \left[\tau_i^1, \tau_i^2, \ldots, \tau_i^{W_i}\right]$	a vector of processing times of task J_i in modes $l_i = 1, 2, \ldots, W_i$,
U	the total amount of a continuous resource available for all tasks J_i at time t, $i = 1, 2, \ldots, n$,
$\underline{u}_i, \overline{u}_i$	given lower and upper bound for the amount of the continuous resource available for task J_i, $i = 1, 2, \ldots, n$,
u_i	the constant amount (unknown in advance) of a continuous renewable resource assigned to task J_i, $i = 1, 2, \ldots, n$,
$u_i(t)$	the amount (unknown in advance) of a continuous renewable resource assigned to task J_i at time t, $u_i(t) \in [0, 1]$, $i = 1, 2, \ldots, n$,
$\boldsymbol{u}(t) = [u_1(t), u_2(t), \ldots, u_n(t)]$	a piecewise continuous (i.e. continuous in a final number of intervals), nonnegative vector function determining the allocation of a continuous resource to the tasks from $\boldsymbol{J}$ at time t,
$\mathbf{u}_i = \left[u_i^1, u_i^2, \ldots, u_i^{W_i}\right]$	a vector of additional resource quantities allocated in each processing mode $l_i = 1, 2, \ldots, W_i$,
W_i	number of available modes of task J_i,
$\tilde{x}_i$	the processing demand (final state) of task J_i, $i = 1, 2, \ldots, n$,
$x_i(t)$	the state of task J_i at time t, $x_i(0) = 0$, $x_i(C_i) = \tilde{x}_i$, $i = 1, 2, \ldots, n$,
$\tilde{x}_{ik}$	a part of processing demand $\tilde{x}_i$ of task J_i corresponding to time interval M_k (combination Z_k), $\tilde{x}_{ik} \ge 0$,
Z_k	the combination of tasks processed in parallel in interval M_k.

8.4 Task Models

8.4.1 Processing Time Versus Resource-Amount Model

Two activity processing models appear in the literature [7]. When continuous resources are taken into account, the processing time versus resource-amount model defines the activity duration as a function of the amount of a continuous resource allocated to this activity. For the case of a single continuous resource this relation can be described as follows:

$$\tau_i = g_i(u_i), \tag{8.1}$$

where $u_i \in [\underline{u}_i, \overline{u}_i]$, and $g_i(u_i) = \infty$ for $u_i \notin [\underline{u}_i, \overline{u}_i]$, $0 < \underline{u}_i \leq \overline{u}_i$.

It is assumed that the resource-amount assigned to J_i does not change during its execution. Within this approach, the existence of some polynomially solvable cases of machine scheduling problems for linear functions is proved in [8].

8.4.2 Processing Rate Versus Resource-Amount Model

In the processing rate versus resource-amount model, the amount of the continuous resource assigned to an activity may change during its processing. The processing rate of an activity is a function of the amount of a continuous resource assigned to this activity at a time. A fundamental result for this model and a renewable resource can be found in [9], whereas for a doubly constrained resource in [10].

As the example of the discussed model, consider the case with a single continuous resource available for all activities at time t in amount U. Activity J_i requires for its processing at time t an amount of a continuous resource available from an interval $[\underline{u}_i, \bar{u}_i]$, where $\underline{u}_i \leq U$, $i = 1, 2, \ldots, n$. The model can be described by an equation:

$$\dot{x}_i(t) = \frac{\mathrm{d}x_i(t)}{\mathrm{d}t} = f_i[u_i(t)], \quad x_i(0) = 0, \quad x_i(C_i) = \tilde{x}_i \tag{8.2}$$

where $f_i[u_i(t)] = \infty$ for $u_i(t) \notin [\underline{u}_i, \overline{u}_i]$, $\underline{u}_i \leq u_i(t) \leq \overline{u}_i$, $0 \leq \underline{u}_i \leq \overline{u}_i$, and equation $\sum_{i=1}^{n} u_i(t) = U$ must be true at time t. The completion of task J_i requires that:

$$x_i(C_i) = \int_0^{C_i} f_i[u_i(t)]\mathrm{d}t = \tilde{x}_i. \tag{8.3}$$

Here, state $x_i(t)$ is an objective measure of work related to the processing of task J_i up to time t. It may denote, for example, the number of man-hours already spent on processing of task J_i, the number of standard instructions in processing of computer program J_i and so on.

At the end of the discussion on activity models, it should be pointed out that the processing rate versus resource-amount model is more natural in the majority of practical situations, since it reflects directly the temporary nature of renewable resources. Moreover, the processing rate versus resource-amount model enables to perform a deeper analysis of the properties of optimal schedules, and can even lead to analytical results in some cases. Because of that, it is sometimes reasonable to treat a discrete resource as a continuous one in order to use this model. Such an approach may be applied when there are sufficiently many allocations of the discrete resource for processing an activity, e.g. in SPP or MPP systems [7].

8.5 Problem Formulation

Discrete-continuous project scheduling problems essentially concerns the allocation of discrete and continuous resources to the activities that are required for their execution. We will consider the DCSP with a single discrete resource, represented by a set of machines, and a single continuous renewable resource. Such problem appears in many practical situations and allows to show the general methodology for discrete-continuous scheduling. The considered problem was studied in [11] for preemptable activities, and in [4] for nonpreemptable, independent activities. The processing rate versus resource-amount model will be used for the task description. The considered DCSP is defined as follows:

Problem P

Let $\boldsymbol{J} = \{J_1, J_2, \ldots, J_n\}$ be a set of independent nonpreemtable tasks, with release dates $r_i = 0$, $i = 1, 2, \ldots, n$, and $\boldsymbol{P} = \{P_1, P_2, \ldots, P_m\}$ be a set of parallel and identical machines, and there is one additional renewable discrete resource. Each task J_i, requires for its processing at time t a machine from $\boldsymbol{P}$ and an amount (unknown in advance) of a continuous renewable resource $u_i(t) \in [0, 1]$, $i = 1, 2, \ldots, n$. Each task from $\boldsymbol{J}$ is described by the processing rate versus resource-amount model. The continuous resource allocation is defined by a piecewise continuous, nonnegative vector function $\boldsymbol{u}(t) = [u_1(t), u_2(t), \ldots, u_n(t)]$, whose values $\boldsymbol{u} = [u_1, u_2, \ldots, u_n]$ are (continuous) resource allocations to the tasks from $\boldsymbol{J}$. It is assumed that the lower—$\underline{u}_i$ and the upper—$\overline{u}_i$ bounds on the amount of the continuous resource $u_i(t)$ to be assigned to J_i at time t for all tasks are the same, i.e. $\underline{u}_i = 0, \overline{u}_i = 1$, $i = 1, 2, \ldots, n$. It is also assumed, without loss of generality, that $\sum_{i=1}^{n} u_i(t) = 1$ for every t.

The problem is to find a sequence of tasks on machines, and, simultaneously, a continuous resource allocation, which minimize given criterion of scheduling Q.

The minimization criteria, among others, might be:

- C_{max}—the makespan (the schedule length), $C_{max} = \max\{C_i\}$,

- $\overline{F}$—the mean flow time $\overline{F} = \frac{1}{n}\sum_{i=1}^{n} F_i$, where $F_i = C_i - r_i$,
- L_{max}—the maximum lateness, $L_{max} = \max\{L_i\}$, $L_i = C_i - d_i$.

An instance **I** of problem **P** is obtained by specifying the values of all its parameters: n, m, d_i, f_i, $\tilde{x}_i$ for $i = 1, 2, \ldots, n$, and the criterion of optimality Q.

According to the definition of the problem it is allowed to allocate $u_i(t) = 0$ to task J_i. Although, this causes the suspension of the task processing, since $f_i(0) = 0$, nevertheless, task J_i still complies with nonpreemtability property. In such a situation, task J_i does not release the machine for other tasks, as it would happen if task was preemtable, and continues processing after some nonzero amount of the continuous resource has been allocated to it.

The defined problem is NP-hard since scheduling nonpreemptable tasks on parallel machines to minimize the schedule length is NP-hard even without any additional resource [12].

8.6 Variants of the DCSP

Now we consider some variants of the discrete-continuous problem **P**. It follows from the definition of **P** that there are no any restrictions on how the continuous resource is allocated to task J_i in the processing rate versus resource-amount model. This way, it is implicitly assumed that the amount of the continuous resource may change during the processing of task J_i. However, there are some practical situations where this assumption is not true [13]. Thus, it would be reasonable to define the class of problems where the amount of the continuous resource is constant during the processing of task J_i. For this reason, two variants ($\mathbf{P_J}$ and $\mathbf{P_M}$) of **P** are defined. In the first variant, it is considered a situation when a constant amount of the continuous resource is allocated to task J_i.

Let $\mathbf{P_J}$ be a discrete-continuous problem **P** where it is additionally assumed that the amount of continuous resource u_i allocated to task J_i is constant during its processing, and $u_i > 0$, for $i = 1, 2, \ldots, n$.

In the next variant, it is considered a situation when a constant amount of the continuous resource is "allocated to machine P_j".

Let $\mathbf{P_M}$ be a discrete-continuous problem $\mathbf{P_J}$ where it is additionally assumed that the amount of continuous resource u_i allocated to tasks J_i assigned to the same machine P_j is the same during their processing.

Because each successive variant of problem **P** arises due to constraints applied to more general preceding problem, among these problems the following dependencies are observed.

Let **I**, $\mathbf{I_J}$, $\mathbf{I_M}$ be the instances of **P**, $\mathbf{P_J}$ and $\mathbf{P_M}$ respectively, and the values of parameters n, m, d_i, f_i, $\tilde{x}_i$ for $i = 1, 2, \ldots, n$ as well as the criterion of optimality Q for all considered instances are the same, then the following two dependences are true [13]:

$$S_{\mathrm{I_M}} \subseteq S_{\mathrm{I_J}} \subseteq S_{\mathrm{I}}, \tag{8.4}$$

where S_{I}, $S_{\mathrm{I_J}}$, and $S_{\mathrm{I_M}}$ are the sets of feasible solutions to the respective DCSP.

If the criterion of optimality Q is minimized, then

$$Q^*(I) \leq Q^*(I_{\mathrm{J}}) \leq Q^*(I_{\mathrm{M}}), \tag{8.5}$$

where Q^*(**I**) is the optimal value of Q for problem instance of the respective DCSP.

8.7 General Approach to Solving the DCSP

The general approach to solving any DCSP assumes decomposition of the problem under consideration into two interrelated sub-problems as follows: (i) construct a feasible sequence of tasks on machines and (ii) allocate the continuous resource among tasks already sequenced [4]. Below we give definitions related to the abovementioned sub-problems.

A feasible schedule for an instance I of a DCSP is a pair ($\mathrm{S_I}$, $\mathrm{D_I}$), $\mathrm{S_I} \in S_I^F$, $\mathrm{D_I} \in D_I^F$, where S_I^F is the set of all feasible sequences of tasks on machines for I, and D_I^F is the set of all feasible continuous resource allocations for established $\mathrm{S_I}$. And $Q(\mathrm{S_I}, \mathrm{D_I})$ will denote the value of scheduling criterion Q for schedule ($\mathrm{S_I}$, $\mathrm{D_I}$).

A semi-optimal schedule for an instance I of a DCSP is an optimal schedule for given $\mathrm{S_I} \in S_I^F$, i.e. a schedule ($\mathrm{S_I}$, $\mathrm{D_I^*}$) for which $Q(\mathrm{S_I}, \mathrm{D_I})$ reaches its minimum:

$$Q^* = \min_{\mathrm{D_I} \in D_I^F} \{Q(\mathrm{S_I}, \mathrm{D_I})\} \tag{8.6}$$

In general, a semi-optimal schedule is obtained by solving the appropriately formulated mathematical programming problem.

An optimal schedule for an instance I of a DCSP is a semi-optimal schedule ($\mathrm{S_I^*}$, $\mathrm{D_I^*}$) for which Q^* reaches its minimum:

$$Q^{**} = \min_{\mathrm{S_I} \in S_I^F} \{Q^*\} \tag{8.7}$$

It follows from the above definitions, that the determining of optimal schedule should be carried out in three steps: firstly, generate a set S^F of all feasible sequences S of tasks on machines, secondly, determine a semi-optimal schedule (S, $\mathrm{D_S^*}$) for each sequence $\mathrm{S} \in S^F$, third, find an optimal schedule (S^*, $\mathrm{D_S^*}$) among all semi-optimal schedules (S, $\mathrm{D_S^*}$).

While the second stage, as it was mentioned above, consists in solving the appropriate mathematical programming problem, the first one, in general, consists in generating all feasible sequences of tasks on machines. Unfortunately, in the general case, the number of feasible sequences $\mathrm{S} \in S^F$ grows exponentially with the number of tasks. Therefore, it would be reasonable to restrict the search space to the smallest subset of S^F such that would contain at least one sequence corresponding to an optimal schedule. We will refer to such set as a *Potentially Optimal Set* (POS). It has been proved in [4], that for concave functions f_i, $i = 1, 2, \ldots, n$, the POS can be reduced to such schedules in which m tasks are executed at every time moment within a schedule. It was also related in [13], that the POS for problem $\mathbf{P_M}$ is formed from such schedules, in which the number of machines not used by the tasks is at most $\max\{m - n, 0\}$.

In the following subsections we will discuss the properties of optimal schedules. Knowledge of these properties is very useful in the construction and analysis of scheduling algorithms, and can even lead to analytical results. We will also consider some particular cases, in which it is possible to reduce the problem of finding optimal schedules for arbitrary concave task models to the solution of some convex programming problems [4].

8.8 Main Properties of Optimal Schedules

In this subsection, we mention only the main properties of optimal schedules that are thoroughly discussed in [4]. Knowledge of these properties is very useful in the construction and analysis of scheduling algorithms, and can even lead to analytical results.

8.8.1 Convex Functions $f_i \leq c_i \cdot u_i$, $c_i = f_i(1)$

For the n task and m machine DCSP with uniform machines and $f_i \leq c_i \cdot u_i$, $c_i = f_i(1)$, $i = 1, 2, \ldots, n$ the makespan C_{max} and the total flow time F are minimized by scheduling all the tasks on the fastest machine (reminder: uniform machines differ by their processing rate, however the rate does not depend on the task carried out). Such schedule implies that the total amount of continuous resource is allocated to a single task at a time, so no other task can be processed in parallel. Thus in this case, in the optimal schedule, tasks are processed sequentially on a single machine [4]. In the optimal schedule that minimizes C_{max} criterion, the order of independent tasks is arbitrary. When the schedule minimizes the total flow time F, the tasks should be ordered according to nondecreasing $\tilde{x}_i/c_i$ values, and for L_{max} criterion, according to the nondecreasing due dates of tasks. The schedules constructed in this way are also optimal for problems $\mathbf{P_J}$ and $\mathbf{P_M}$. Although, optimal schedules of independent tasks with convex processing rates are easy to construct ($O(n\log n)$ [13]), this case has no practical importance, since convex functions f_i do not appear in practice.

8.8.2 Concave Functions f_i and $n \leq m$

In this particular case, the task sequencing problem does not arise at all, and it is justified to consider the case $n = m$, since all n tasks can be assigned to any of m identical machines, and $m - n$ machines will remain idle. Thus, the only problem, that has to be solved, is the allocation of continuous resource to tasks with concave processing rate functions. In such case, according to [4], the makespan is minimized by fully parallel processing of all tasks using the following resource-amounts:

$$u_i^* = f_i^{-1}(\tilde{x}_i/M^*), \quad i = 1, 2, \ldots, n, \tag{8.8}$$

where M^* is the (unique) positive root of the equation:

$$\sum_{i=1}^{n} f_i^{-1}(\tilde{x}_i/M^*) = 1. \tag{8.9}$$

As it can be seen, the amount of continuous resource allocated to the tasks is constant, which makes the obtained schedule optimal also for $\mathbf{P_J}$ and $\mathbf{P_M}$ with minimization criteria C_{max} and F [13].

The considered case is important from the practical point of view, because in some particular cases it is possible to find an optimal solution analytically. This case is also of fundamental importance for the general methodology for solving any DCSP with $n > m$, including the ones with precedence constraints among tasks.

8.8.3 Concave Functions f_i and $n > m$

The methodology for dealing with this case of the DCSP is based on the findings for the case $n \leq m$. First of all, the notion of feasible sequence of tasks S^z is introduced. Each feasible schedule (i.e. a feasible solution of a discrete-continuous problem) can be divided into $p \leq n$ intervals of length M_k, $k = 1, 2, \ldots, p$, defined by the completion times of consecutive tasks. Let Z_k denote the combination of tasks processed in parallel in the interval M_k. Thus, in general, a feasible sequence S^z of combinations Z_k, $k = 1, 2, \ldots, p$ can be associated with each feasible schedule. Feasibility of such sequence requires, in addition to the number of elements in each combination restricted by m, that each task appears in at least one combination, and that nonpreemptability of each task is guaranteed. The last condition means, that each task appears exactly in one or in consecutive combinations in S^z. The presented idea is illustrated by an example in Fig. 8.1.

In this example, a feasible schedule of seven tasks on three machines is considered. It is assumed for simplicity, that the same amount of continuous resource is allocated to each task, i.e. $u_i(t) = 1/3$ for every t. The feasible sequence S^z of

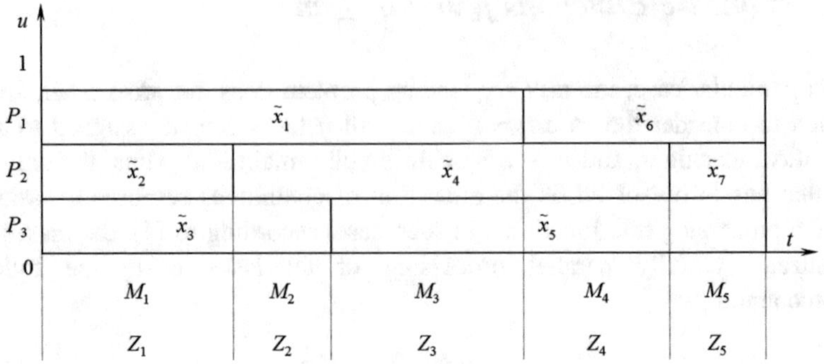

Fig. 8.1 The division of a feasible schedule into intervals M_k defined by the completion times of consecutive tasks

combinations Z_k, corresponding to the considered schedule is: $S^z = \{1, 2, 3\}$, $\{1, 4, 3\}$, $\{1, 4, 5\}$, $\{6, 4, 5\}$, $\{6, 7, 5\}$. Further, the processing demand of each task can be divided into parts $\tilde{x}_{ik} \geq 0$ corresponding to particular time intervals M_k (combinations Z_k), see Fig. 8.2.

For a given feasible sequence of tasks on machines one can find an optimal division of processing demands of tasks $\tilde{x}_i$, $i = 1, 2, \ldots, n$, among combinations $Z_k \in S^z$, i.e. a division which leads to a minimum length schedule from among all feasible schedules generated by S^z. Once the division is known, the optimal continuous resource allocation to the task parts $\tilde{x}_{ik}$ within a particular combination Z_k can be determined in the same way as in the case $n = m$. Also, on the basis of the results for the case $n = m$, the length of interval M_k can be calculated as the unique positive root of equation:

$$\sum_{i \in Z_k} f_i^{-1}(\tilde{x}_{ik}/M_k^*) = 1. \tag{8.10}$$

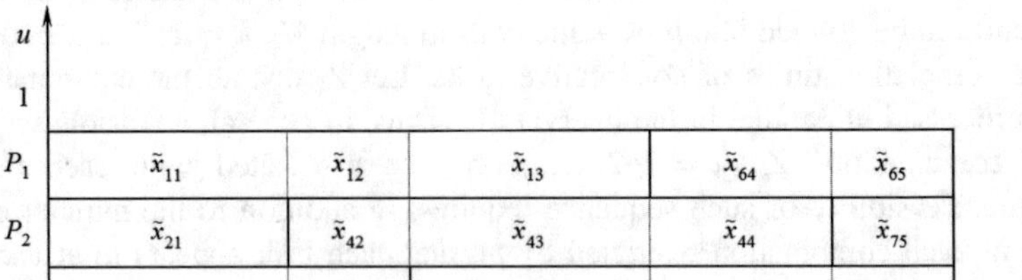

Fig. 8.2 The division of processing demands of tasks $\tilde{x}_i$ into parts $\tilde{x}_{ik}$, corresponding to time intervals M_k

Having calculated M_k^*, the amount of continuous resource allocated to task J_i within k-th interval is calculated as:

$$u_{ik}^* = f_i^{-1}(\tilde{x}_{ik}/M_k^*), \quad i \in Z_k. \tag{8.11}$$

It should be mentioned that the obtained amount of continuous resource u_{ik}^* allocated to $\tilde{x}_{ik}$ remains constant within interval k, $k = 1, 2, \ldots, p$ [4]. Function $\boldsymbol{u}(t) = [u_1(t), u_2(t), \ldots, u_n(t)]$, $t \in [0, C_{max}]$ takes at most p different values, dependent on the division of the processing demands $\tilde{x}_i$ among combinations Z_k. These values can change only at the time moments defined by the completion times of the consecutive tasks. It has been shown in [14] that the value of each considered criteria, C_{max}, L_{max}, and $\overline{F}$ depends on the length of intervals M_k, $k = 1, 2, \ldots, p$.

The above described approach for solving the continuous resource allocation problem through the division of processing demands of tasks can be used for finding a semi-optimal schedule $(S^z, D_{S^z}^*$ for the considered case of concave functions f_i and $n > m$. For this purpose a convex mathematical programming problem $\Omega(C_{max})$ dealing with optimal division of the processing demands $\tilde{x}_i$ among combinations Z_k's has to be solved [4]:

Problem $\Omega(C_{max})$

$$\min \quad C_{\max} = \sum_{k=1}^{p} M_k^*(\{\tilde{x}_{ik}\}_{i \in Z_k}), \tag{8.12}$$

$$\text{s.t.} \quad \sum_{k \in K_i} \tilde{x}_{ik} = \tilde{x}_i, \quad i = 1, 2, \ldots, n, \tag{8.13}$$

$$\tilde{x}_{ik} \geq 0, \quad i = 1, 2, \ldots, n, \quad k \in K_i, \tag{8.14}$$

where K_i - the set of all indices of Z_k's such that task $J_i \in Z_k$, and $M_k^*\left(\{\tilde{x}_{ik}\}_{i \in Z_k}\right)$ is calculated from Eq. (8.10) for $k = 1, 2, \ldots, p$.

The mathematical programming problem obtained is always a convex one, because a sum of convex functions is also convex. In this optimization problem, the sum of the minimum-length intervals (i.e. parts of a feasible schedule) generated by consecutive combinations in S^z, as functions of the $\tilde{x}_{ik}$'s, is minimized subject to the constraints that each task has to be completed. In consequence, an optimal schedule can be found by solving the continuous resource allocation problem optimally for all feasible sequences S^z from the POS. It has been proved in [4], that for concave functions f_i, $i = 1, 2, \ldots, n$, a set containing all feasible sequences composed of $p = n - m + 1$ m-element combinations of tasks Z_k, $k = 1, 2, \ldots, p$, comprises the POS.

8.8.3.1 Identical Concave Functions

It was proved in [14] that for the case with identical concave functions $f_i = f(u_i)$, the problem of determining a semi-optimal schedule can be simplified to solving a single nonlinear equation. Namely, for a given feasible sequence S^z such that $|Z_k| = m$ for each $k = 1, 2, \ldots, n - m + 1$ an optimal continuous resource allocation can be determined by solving the equation:

$$\sum_{j=1}^{m} f_i^{-1}(\tilde{x}_j / C^*_{\max}) = 1, \tag{8.15}$$

where

$$\tilde{x}_j = \sum_{i \in J_j} \tilde{x}_i, \tag{8.16}$$

and J_j is the set of indices of tasks assigned to machine P_j, $j = 1, 2, \ldots, m$ as determined by feasible sequence S^z.

In the considered case, all tasks carried out on the same machine use the same constant amount of continuous resource which, in general, is different across the machines. Moreover, this case meets the assumptions of problem $\mathbf{P_M}$, whose POS(C_{max}) is significantly smaller than the POS of problem $\mathbf{P}$. This would justify solving problem $\mathbf{P_M}$ instead of $\mathbf{P}$ in this case [13].

Note the special case, when scheduling of tasks is performed only on two machines. When the task processing rate functions are strictly concave, then problem $\mathbf{P}$ can be solved in pseudo-polynomial time [13]. In this case, minimization of makespan M is equivalent to minimization of a function $C_{max} = \max\{\tilde{x}_1, \tilde{x}_2\}$, where $\tilde{x}_1$ and $\tilde{x}_2$ are the total processing demands of tasks assigned to the respective machines. This case can be solved optimally in pseudo-polynomial time using dynamic programming [15].

8.8.3.2 Concave Power Functions

Power task processing rate functions is a special class of the DCSP that covers a wide range of practical applications. Before we move on to the particular cases of concave power functions, two general for this class of functions properties are recalled.

The first property (**p-1**) claims, that the DCSP with task processing demands $\tilde{x}_1, \ldots, \tilde{x}_n$ and processing rates $f_i = c_i u_i^{1/\alpha_i}$, $\alpha_i \geq 1$, $c_i > 1$, $i = 1, 2, \ldots, n$ is equivalent to the DCSP with task processing demands $\tilde{y}_1, \ldots, \tilde{y}_n$ and processing rates $f_i = u_i^{1/\alpha_i}$, $\alpha_i \geq 1$, $i = 1, 2, \ldots, n$, where $\tilde{y}_i = \tilde{x}_i / c_i$ (proved in [16] for the makespan and in [17] for the mean flow time as the optimality criteria).

The second property (**p-2**) claims, that for the DCSP with task processing rates $f_i = u_i^{1/\alpha_i}$, $\alpha_i \geq 1$, $i = 1, 2, \ldots, n$ there exists an optimal schedule in which at most one task with $\alpha_i = 1$ occurs in each combination Z_k of the corresponding feasible sequence S^z. Moreover, if the number of tasks with $\alpha_i \geq 2$ is not less than m, then in any optimal schedule each combination Z_k of the corresponding feasible sequence S^z contains at most one task with $\alpha_i = 1$ [16]. This property allows to reduce the POS only to those sequences S^z which comply with it and shorten the process of obtaining the optimal schedule, since there is no need to calculate optimal allocations of continuous resource to the sequences which do not comply with the property.

Below, the most important from the practical point of view cases of the power functions are recalled from [16].

i. *Concave Power Functions* $f_i = u_i^{1/\alpha_i}$, $\alpha_i \in \{1, 2, 3, 4\}$, $i = 1, 2, \ldots, n$

In this case, Eq. (8.10) is reduced to its algebraic form of the power ≤ 4 and thereby can be solved analytically. This property significantly facilitates the determination of the optimal assignment of continuous resource to the tasks from feasible sequence S^z, i.e. the resolution of problem $\mathbf{\Omega}(C_{max})$. These functions are especially important, because allow to model task processing rates in various practical problems [4].

ii. *Concave Identical Power Functions* $f_i = u_i^{1/\alpha}$, $\alpha > 1$, $i = 1, 2, \ldots, n$

Since the power functions are identical, the general property for the case of identical concave functions defined in Sect. 8.8.3.1 can be applied to the considered special case. This property simplifies the problem $\mathbf{\Omega}(C_{max})$ of determining a semi-optimal schedule $(S^z, D^*_{Sz}$ for given S^z to solving only one nonlinear equation given below:

$$C^*_{\max} = \left(\sum_{j=1}^{m} \left(\sum_{i \in J_j} \tilde{x}_i \right)^{\alpha} \right)^{1/\alpha}, \tag{8.17}$$

where J_j is the set of indices of tasks assigned to machine P_j, $j = 1, 2, \ldots, m$ as determined by feasible sequence S^z and $\alpha > 1$.

iii. *Concave Power Functions* $f_i = u_i^{1/\alpha_i}$, $\alpha_i \in \{1, 2\}$, $i = 1, 2, \ldots, n$

For this case of power functions, two subsets of tasks are distinguished:

- A_1 — the subset of n_1 tasks with power functions $f_i = u_i$, i.e. $\alpha_i = 1$, and
- A_2 — the subset of n_2 tasks with power functions $f_i = u_i^{1/2}$, i.e. $\alpha_i = 2$, and $n_1 + n_2 = n$.

Using the above division of tasks, four special sub-cases are identified:

$\boldsymbol{n_1 = n, n_2 = 0}$
In this sub-case all tasks have the same processing rates $f_i = u_i$ whose values linearly depend on the amount of continuous resource allocated to particular tasks. It follows from Eq. (8.17) that for such processing rates any assignment of tasks to identical machines results in the same value of the makespan, recall also the property from Sect. 8.8.1.

$\boldsymbol{n_1 = 0,\ n_2 = n}$
Since there are only tasks with processing rates $f_i = u_i^{1/2}$ and no tasks with $f_i = u_i$, the considered case complies to subcase (ii). Therefore, the length of semi-optimal schedule can be determined using Eq. (8.17).

$\boldsymbol{n_1 > 0,\ n_2 < m}$
When the number of tasks n_2 with processing rate functions $f_i = u_i^{1/2}$ is less than the number of machines m, an optimal schedule can be obtained using property **p-2**. All tasks from set A_2 are to be scheduled on different machines and all tasks from set A_1 are to be scheduled on the same still unused machine. Notice that in this case $m - n_2 - 1$ machines remain idle.

$\boldsymbol{n_1 > 0,\ n_2 \geq m}$
This sub-case with task processing rates $f_i = u_i^{1/\alpha_i}$, $\alpha_i \in \{1, 2\}$, $i = 1, 2, \ldots, n$ is completely consistent with property **p-2** where the same processing rates are considered and $\alpha_i \geq 1$. Therefore, if feasible sequence S^z insures that at most one task with $\alpha_i = 1$ is processed at any time moment within the schedule, then the problem $\boldsymbol{\Omega}(C_{\max})$ of determining a semi-optimal schedule $(S^z, D^*_{S^z}$ for given S^z can be solved analytically using the equation below:

$$C^*_{\max} = \frac{1}{2}\left(\sum_{J_i \in A_1} \tilde{x}_i' + \sqrt{\sum_{j=1}^{m}\left(\sum_{i \in J_j} \tilde{x}_i'\right)^2}\right), \tag{8.18}$$

where J_j is the set of indices of tasks assigned to machine P_j, $j = 1, 2, \ldots, m$ as determined by feasible sequence S^z and

$$\tilde{x}_i' = \begin{cases} \tilde{x}_i & \text{if} \quad \alpha_i = 1, \\ 2\tilde{x}_i & \text{if} \quad \alpha_i = 2. \end{cases} \tag{8.19}$$

8.9 Minimization of the Maximum Lateness $L_{\max}$

The problem of minimization of the maximum lateness $\boldsymbol{\Omega}(L_{\max})$ defined in [18] is recalled in this section. For the purpose of definition it was assumed, that the index of a task corresponds to the number of the last combination $Z_k \in S^z$ in which this task appears.

Problem $\boldsymbol{\Omega}(L_{\max})$

$$\min \quad L_{\max} = \max\{L_i\}, L_i = C_i - d_i, \tag{8.20}$$

$$\text{s.t.} \quad \sum_{k \in K_i} \tilde{x}_{ik} = \tilde{x}_i, \quad i = 1, 2, \ldots, n, \tag{8.21}$$

$$\sum_{j \le k} M_j^*(\{\tilde{x}_{ij}\}_{i \in Z_j}) - d_k \le L_{\max}, \quad k = 1, 2, \ldots, n, \tag{8.22}$$

$$\tilde{x}_{ik} \ge 0, \quad i = 1, 2, \ldots, n, \quad k \in K_i, \tag{8.23}$$

where K_i - the set of all indices of Z_k's such that task $i \in Z_k$, and $M_j^*\left(\{\tilde{x}_{ij}\}_{i \in Z_j}\right)$ is calculated from Eq. (8.10).

Constraints 8.21 ensure the execution of all tasks, and constraints 8.22 arise during the transformation of minimax object function into a linear one.

In the general case, for problem **P** with independent tasks and concave processing rate functions f_i, $i = 1, 2, \ldots, n$, POS($L_{\max}$) contains feasible sequences S^z consisting of n combinations Z_k. The first $p = n - m + 1$ combinations Z_k contain m tasks, and combinations $n - m + 2, \ldots, n - 1, n$ contain $m - 1, \ldots, 2, 1$ tasks respectively.

When there are precedence constraints among the tasks, POS($L_{\max}$) also contains feasible sequences S^z consisting of n combinations Z_k. However in this case, such sequences S^z are allowed for which the number of tasks in combinations Z_k is not less than 1 and is greater than the number of tasks in this combination for the case of independent tasks. For this reason, the cardinality of POS($L_{\max}$) in such case is much greater than in case with independent tasks.

Unfortunately, in both considered cases independently on the existence of precedence constraints among the tasks the cardinality of POS($L_{\max}$) grows exponentially with the number of tasks n.

8.10 Minimization of Mean Flow Time $\overline{F}$

The problem of minimization of mean flow time $\boldsymbol{\Omega}(\overline{F})$ defined in [17] and later developed in [14] and [19] is recalled in this section.

According to the formulation of problem **P**, the release dates of tasks $r_i = 0$, $i = 1, 2, \ldots, n$, and the respective objective function for the considered optimization criterion is described by equation $\overline{F} = \frac{1}{n}\sum_{i=1}^{n} F_i$, where $F_i = C_i - r_i$. An optimal allocation of the continuous resource to given feasible sequence S^z can be determined by solving the respective convex mathematical programming problem:

Problem $\Omega(\overline{F})$

$$\min \quad \overline{F} = \frac{1}{n}\sum_{k=1}^{p} (n-k+1)M_k^*(\{\tilde{x}_{ik}\}_{i\in Z_k}), \tag{8.24}$$

$$\text{s.t.} \quad \sum_{k\in K_i} \tilde{x}_{ik} = \tilde{x}_i, \quad i = 1, 2, \ldots, n, \tag{8.25}$$

$$\tilde{x}_{ik} \geq 0, \quad i = 1, 2, \ldots, n, \quad k \in K_i, \tag{8.26}$$

where K_i - the set of all indices of Z_k's such that task $i \in Z_k$, and $M_k^*\left(\{\tilde{x}_{ik}\}_{i\in Z_k}\right)$ is calculated from Eq. (8.10). In the formulation of the problem, Eq. (8.24) is the weighted sum of minimal lengths intervals M_k^* corresponding to subsequent combinations Z_k in S^z, which is a convex function. And constraints (8.25) ensure the execution of all tasks. For this problem, a feasible solution can be represented by a feasible sequence of n combinations of tasks, such that first $n - m + 1$ combinations contained exactly m elements and the consecutive combinations $n - m + 2$, $n - m + 3, \ldots, n - 1, n$ contained $m - 1, m - 2, \ldots, 1$ element, respectively.

Unfortunately, in the general case, for problem **P** with and without precedence constraints among the tasks and concave processing rate functions f_i, $i = 1, 2, \ldots, n$, POS($\overline{F}$) is constructed similarly as POS(L_{max}). Hence, the cardinality of POS($\overline{F}$) also grows exponentially with the number of tasks n.

In [19], a problem of the total completion time minimization was considered and some new properties of optimal schedules were presented. There were proposed two theorems which allow to determine the continuous resource allocation analytically. These findings can also be applied to solve the continuous part of the mean flow minimization problem, since both problems are equivalent.

The first theorem states, that for a discrete-continuous scheduling problem with processing rates of tasks $f_i(u_i) = c_i u_i^{1/\alpha_i}$, $\alpha_i \geq 1$, $i = 1, 2, \ldots, n$, an optimal resource allocation (for a given feasible sequence) with respect to the mean flow time requires finding roots of a polynomial of order at most $\max\{\alpha_i\}$. Thus, for $\alpha_i \in \{1, 2, 3, 4\}$ the optimal resource allocation can be found analytically in polynomial time.

The second theorem states, that for a discrete-continuous scheduling problem with processing rates of tasks $f_i(u_i) = c_i u_i^{1/\alpha_i}$, $\alpha_i \geq 1$, $i = 1, 2, \ldots, n$, there exists a mean flow time optimal schedule such that at most one task with $\alpha_i = 1$ is scheduled in each interval $[C_{i-1}, C_i]$, $i = 1, 2, \ldots, n$, $C_0 = 0$. Moreover, if $\alpha_i = \alpha_j$, and $\tilde{x}_i \leq \tilde{x}_j$ then $C_i \leq C_j$.

The two cited above theorems can be used to improve the efficiency of the local search algorithms, which can be developed for the considered problem, either by substituting the time-consuming step of solving the mathematical programming problem for the continuous resource allocation, or by reducing the search space.

References

1. Hartmann, S., Briskorn, D.: A survey of variants and extensions of the resource-constrained project scheduling problem. Eur. J. Oper. Res. **207**(1), 1–14 (2010)
2. Błazewicz, J., Lenstra, K., Rinnooy Kan, A.H.G.: Scheduling subject to resource constraints: classification and complexity. Discret. Appl. Math. **5**(1), 11–24 (1983)
3. Węglarz, J., Blazewicz, J., Cellary, W., Słowiński, R.: Algorithm 520: an automatic revised simplex method for constrained resource network scheduling. ACM Trans. Math. Softw. **3**, 295–300 (1977)
4. Józefowska, J., Węglarz, J.: On a methodology for discrete-continuous scheduling. Eur. J. Oper. Res. **107**(2), 338–353 (1998)
5. Janiak, A.: Minimization of the blooming mill standstills—mathematical model. Suboptimal algorithms. Zesz. Nauk. AGH, s. Mechanika **8**(2), 37–49 (1989)
6. Węglarz, J.: Multiprocessor scheduling with memory allocation—a deterministic approach. IEEE Trans. Comput. **C29**(8), 703–709 (1980)
7. Węglarz, J., Józefowska, J., Mika, M., Waligóra, G.: Project scheduling with finite or infinite number of activity processing modes—a survey. Eur. J. Oper. Res. **208**, 177–205 (2011)
8. Janiak, A.: Single machine scheduling problem with a common deadline and resource dependent release dates. Eur. J. Oper. Res. **53**(3), 317–325 (1991)
9. Węglarz, J.: Time-optimal control of resource allocation in a complex of operations framework. IEEE Trans. Syst. Man Cybern. **6**(11), 783–788 (1976)
10. Węglarz, J.: Project scheduling with continuously-divisible, doubly constrained resources. Manage. Sci. **27**(9), 1040–1052 (1981)
11. Węglarz, J.: Project scheduling with discrete and continuous resources. IEEE Trans. Syst. Man Cybern. **9**(10), 644–650 (1979)
12. Józefowska, J., Różycki, R., Waligóra, G., Węglarz, J.: Local search metaheuristics for discrete-continuous scheduling problems. Eur. J. Oper. Res. **107**(2), 354–370 (1998)
13. Różycki, R.: Zastosowanie algorytmu genetycznego do rozwiązywania dyskretno-ciągłych problemów szeregowania. Ph.D. Dissertation, Poznań University of Technology, Poland (2000)
14. Józefowska, J., Mika, M., Różycki, R., Waligóra, G., Węglarz, J.: Discrete-continuous scheduling to minimize the mean flow time—computational experiments. Comput. Methods Sci. Technol. **3**(1), 25–37 (1997)
15. Blazewicz, J., Kubiak, W., Szwarcfiter, J.: Scheduling independent fixed-type tasks. In: Słowiński, R., Węglarz, J. (eds.) Advances in Project Scheduling. Elsevier, Amsterdam (1989)
16. Józefowska, J., Mika, M., Różycki, R., Waligóra, G., Węglarz, J.: Discrete-continuous scheduling to minimize the makespan for power processing rates of jobs. Discret. Appl. Math. **94**, 263–285 (1999)
17. Józefowska, J., Węglarz, J.: Discrete-continuous scheduling problems—mean completion time results. J. Oper. Res. **94**, 302–309 (1996)

18. Józefowska, J., Mika, M., Różycki, R., Waligóra, G., Węglarz, J.: Minimalizacja maksymalnego opóźnienia w dyskretno-ciągłych problemach szeregowania—algorytmy heurystyczne. Zeszyty Naukowe Politechniki Śląskiej, Seria: Automatyka z. **123**, 221–231 (1998)
19. Józefowska, J., Węglarz, J.: New results for discrete-continuous mean flow time scheduling problems. In: Eight International Workshop on Project Management and Scheduling, Valencia, Spain, pp. 217–220 (2002)

Chapter 9
State-of-the-Art Review

9.1 Theoretical Research on the DCSP

As it is known from Sect. 8.7, the general methodology for solving any DCSP assumes firstly, determining feasible sequences of tasks on machines, and secondly, determining optimal allocation of continuous resource to these sequences. The second stage of solving the DCSP is formulated as a convex mathematical programming problem with linear constraints which can be solved using the appropriate solver. Thus, to put it more precisely, solving any DCSP assumes decomposition of the DCSP into two interrelated sub-problems: (i) construct a feasible sequence of tasks on machines and (ii) allocate the continuous resource among tasks already sequenced [1]. In order to determine the optimal allocation of continuous resource to a given sequence of tasks on machines, a convex mathematical programming problem has to be formulated and solved. This problem can be solved either analytically, or by approximation of the function value. Unfortunately, the problem is not always easy to solve analytically even for simple cases, and there are cases for which there is no analytical solution [2], and the approximation of the function value using specialized solvers can be very time-consuming. The difficulty of the problem motivated theoretical research on finding the ways to simplify the solution of the problem.

9.1.1 Another Formulation of the DCSP

In [2], another formulation of the problem was proposed, where time M_k (the time of processing the k-th subset of tasks parts) was given explicitly as optimized variable t_k. In order to achieve this, the penalty function idea was used to construct a convex objective function for the k-th subset of tasks parts, which reaches its unique minimum at point T_k. Thus, the objective function is given as follows:

E. Ratajczak-Ropel and A. Skakovski, *Population-Based Approaches to the Resource-Constrained and Discrete-Continuous Scheduling*,
Studies in Systems, Decision and Control 108, DOI 10.1007/978-3-319-62893-6_9

$$g(t_k, \tilde{\mathbf{x}}_k) \triangleq t_k + \phi(t_k, \tilde{\mathbf{x}}_k), \tag{9.1}$$

where vector $\tilde{\mathbf{x}}_k \triangleq (\{\tilde{x}_{ik}: i \in Z_k\})$. The expanded definition of the function is given in [2] or [3]. Having the above function for each subset Z_k, the authors formulated an optimization problem $\Omega^A(C_{max})$ equivalent to problem $\Omega(C_{max})$ (see Eqs. (8.12)–(8.14) in Sect. 8.8.3), in which the values $\tilde{\mathbf{x}}_k$, $k = 1, 2, \ldots, p$, of each solution are equal to the solution of problem $\Omega(C_{max})$. Thus, optimization problem $\Omega^A(C_{max})$ is defined as follows:

$$\min \quad C_{\max} = \sum_{k=1}^{p} g(t_k, \tilde{\mathbf{x}}_k), \tag{9.2}$$

$$\text{s.t.} \quad \sum_{k \in K_i} \tilde{x}_{ik} = \tilde{x}_i, \quad i = 1, 2, \ldots, n, \tag{9.3}$$

$$x_{ik} \geq 0, \quad k = 1, 2, \ldots, p, \; i \in Z_k \tag{9.4}$$

$$t_k \geq 0, \quad k = 1, 2, \ldots, p, \tag{9.5}$$

where K_i - the set of all indices of Z_k's such that task $i \in Z_k$. The authors point out two important benefits of such formulation. First, there is no need to solve Eq. (8.10) to construct objective function in order to solve the optimization problem. Second, such objective function is convex as the sum of convex functions $g(t_k, \tilde{\mathbf{x}}_k)$. Unfortunately, there are also disadvantages, most important of which is that the objective function does not have continuous derivative and this makes it inefficient for methods based upon functions with continuous derivatives. Thus, additional computation is needed to smooth the function and this makes the whole procedure slow. Another disadvantage is that the number of optimization variables increases by p, from $p \cdot (m + 1)$. To overcome these disadvantages, a new approach was proposed in [3], where the number of optimization variables remains $p \cdot (m + 1)$, but the objective function is linear and problem can be solved efficiently. The discussion on the proposed new approach and a comparative experiment involving $\Omega(C_{max})$ and $\Omega^A(C_{max})$ is provided in the next Section.

9.1.2 The New Approach to Optimal Resource Allocation

In [3], based on their theoretical findings discussed in [4, 5], the authors proposed the new approach to optimal continuous resource allocation. This new approach is based on the findings on the problem inverse to the considered one, i.e. the problem to minimize the use of the continuous resource required to carry out the tasks within a given upper bound on makespan $\widehat{C}$, (this problem will be referred to as $\Omega(U_{\max})$ in the rest of the discussion). So, it is proposed by the authors to solve problem

$\Omega(C_{max})$ by solving $\Omega(U_{\max})$ for the upper bound on the schedule length equal to optimal schedule length. Thereby, the original optimization problem $\Omega(C_{max})$ can be formulate as problem $\Omega^{\mathrm{N}}(C_{max})$ as follows:

$$\min \quad C_{\max} = \sum_{k=1}^{p} t_k, \tag{9.6}$$

$$\text{s.t.} \quad \sum_{i \in Z_k} f_i^{-1}\left(\frac{\tilde{x}_{ik}}{t_k}\right) = \widehat{U}, \quad k = 1, 2, \ldots, p, \tag{9.7}$$

$$\sum_{k \in K_i} \tilde{x}_{ik} = \tilde{x}_i, \quad i = 1, 2, \ldots, n, \tag{9.8}$$

$$x_{ik} \geq 0, \quad k = 1, 2, \ldots, p, \quad i \in Z_k \tag{9.9}$$

$$t_k \geq 0, \quad k = 1, 2, \ldots, p, \tag{9.10}$$

According to this formulation, the value of M_k for given task parts $\tilde{x}_{ik}$ can be obtained by explicitly optimizing variable t_k which corresponds to the time of processing parts $\tilde{x}_{ik}$ of tasks from subset Z_k, subject to the constraint on resource level Eq. (9.7). The solution to the considered problem is the optimal resource allocation for a given feasible sequence S^z. Moreover, now the objective function is the linear one.

An experiment was carried out to compare the performance of Sequential Quadratic Programming (SQP) method, provided by MATLAB Optimization Toolbox as `fmincon` procedure, applied for solving problems $\Omega(C_{max})$, $\Omega^{\mathrm{A}}(C_{max})$, and $\Omega^{\mathrm{N}}(C_{max})$. MATLAB's `fmincon` procedure is reported to be one of the best in constrained nonlinear programming [6]. In order to obtain objective function of problem $\Omega(C_{max})$, both, symbolic solution (to solve Eq. (8.10) and construct the objective function before optimization starts), as well as approximation method, performed each time when SQP procedure had to calculate the value of the objective function, were used. In the experiment, the concave processing rate functions of the form $f_i(u_i) = c_i \cdot u_i^{1/\alpha_i}$ were used. For this functions with α_i $\{1, 2, 3, 4\}$ Eq. (8.10) can be solved analytically. The number of machines m = 2, 3, 4, 5, 6 and the number of tasks for the sets $\alpha_i \in \{1, 2\}$ and $\alpha_i \in \{1, 2, 3\}$ was equal 20. Because both symbolic and optimizing calculation was very time consuming for problem $\Omega(C_{max})$, the number of tasks for $\alpha_i \in \{1, 2, 3, 4\}$ was reduced to 10 (for m = 6 the size of the file with the objective function had a few megabytes). The SQP procedure was used with the default settings and in most cases stopped after performing the default number of iterations equal $100 \times$ number of variables, before reaching the default precision. The performance of the SQP procedure applied for solving $\Omega(C_{max})$, $\Omega^{\mathrm{A}}(C_{max})$, and $\Omega^{\mathrm{N}}(C_{max})$ was compared using mean relative error δ, calculated as:

$$\delta = \frac{M' - M^*}{M'} \cdot 100\%, \tag{9.11}$$

where M' is the solution to problem $\Omega(C_{max})$ obtained using symbolically generated objective function, and M^* is the solution obtained for $\Omega(C_{max})$ using approximation instead, or the solution to $\Omega^A(C_{max})$ or $\Omega^N(C_{max})$. The mean relative errors of the solutions found by the SQP procedure for each set of 20 instances was similar, and did not exceed 3%. Oppositely, the difference in time, required to yield a solution was significant and to great extent depended on the problem and its size. In case of $\Omega(C_{max})$ and $\alpha_i \in \{1, 2, 3, 4\}$ solved symbolically, the time to solve the problem increased rapidly with the growth of the number of machines from 0,4 s to 193,8 s. Similarly, when $\Omega(C_{max})$ was solved by the approximation of objective function, the time required to yield a solution was several orders of magnitude greater than for any other optimization problem under the test. It took from 127,4 s to 4455,4 s to find a solution in this case. On the contrary, the time required to yield a solution for $\Omega^A(C_{max})$ and $\Omega^N(C_{max})$ was respectively 3,8 s–92,5 s and 1,1 s–16,1 s. All experiments were carried out on PC computer with Intel Xeon X3220 2, 4 GHz processor and 4 GB RAM memory in MATLAB environment. The main observations of the experiment are such that in the case of simple instances, the time required for the SQP to solve $\Omega^N(C_{max})$ was greater than the time required to solve $\Omega(C_{max})$ with the symbolic objective function. In the case of instances of medium difficulty, it was similar to the time required to solve $\Omega(C_{max})$ symbolically and significantly better in the case of difficult instances ($\Omega(C_{max})$: 0,8 s–193,8 s, $\Omega^N(C_{max})$: 1,1 s–1,6 s).

9.1.3 New Properties of the Discrete Part of the DCSP

In [7, 8], a theoretical research on the discrete part of the DCSP was conducted. In [7], the authors introduced properties which enabled the reduction of the solution space of the discrete part of the DCSP with concave processing rate functions. A special subset T_1 of tasks $\boldsymbol{J}$, containing all tasks which appear only in one combination $Z_k \in S^z$, was used to distinguish all such combinations $Z_k \in S^z$ which contain a task from T_1, and, thereby, such combinations Z_k which do not contain a task from T_1. This distinction gave a possibility to identify subsequences $S_{k,l} \in S^z$ of consecutively numbered combinations Z_k which contain a task from T_1. Based on these assumptions, it was proved in [7] that a subset $\widehat{S}$ of the set of all feasible sequences $\boldsymbol{S}^z$ contains an optimal solution to the considered problem provided that the numbers of tasks from T_1 appear in each $S_{k,l} \in S^z$, $S^z \in \boldsymbol{S}^z$ in an increasing order. Next, subset $\widehat{S}$ is partitioned into two subsets: $\widehat{S}_0$, consisting of such feasible sequences S^z in which each Z_k contains a task from T_1, and $\widehat{S}_1$, composed of feasible sequences S^z in which at least one combination Z_k does not contain a task from T_1. This way distinguished subsets of $\widehat{S}$, helped to prove several properties [7] which made it possible to reduce the solution space and evaluate its cardinality. The reduction of the solution space determined by $\widehat{S}$ was obtained by the removal of the equivalent, in terms of objective function value, sequences from $\widehat{S}$. This way, subset $\widehat{S}$ was reduced to subset $\widehat{S^*}$ which was proved [8] to be the largest subset of $\widehat{S}$ which does not contain equivalent sequences and contain an optimal sequence S^z. It was also proved in [8], that the cardinality of subset $\widehat{S^*}$ can be evaluated with regard to the number of tasks n and machines m by the following equation:

$$\binom{n}{n-m+1}+\frac{1}{2}\cdot\sum_{r=1}^{n-m-1}\binom{n}{m-1+r}\cdot\binom{m-1+r}{m-1}\cdot r!\cdot(m-1)^r\cdot \sum_{i=1}^{n-m-r}\binom{n-m+1-r}{i}+\sum_{r=2}^{n-m-1}\binom{n}{m-1+r}\cdot\binom{m-1+r}{m-1}\cdot r!\cdot(m-1)^r\cdot \sum_{i=1}^{n-m-1-r}\binom{n-m+1-r}{i}\cdot\sum_{j=1}^{n-m-r-i}\binom{n-m+1-r-i}{j}\cdot(r-1)^{n-m+1-r-i-j}. \tag{9.12}$$

9.2 Discretisation of the DCSP

According to the general approach introduced in Sect. 8.7, solving any DCSP assumes decomposition of the DCSP into two interrelated sub-problems: (i) construct a feasible sequence of tasks on machines and (ii) allocate the continuous resource among tasks already sequenced [1]. In order to determine the optimal allocation of continuous resource to a given sequence of tasks on machines, and therefore a semi-optimal schedule (S, D_S^*), a convex mathematical programming problem has to be formulated and solved. Solution of this problem gives the schedule of the minimal length among the infinite number of schedules generated by given sequence of tasks on machines S. Unfortunately, finding the optimal solution is hardly suitable for the practice, because mathematical programming problems are, in general, computationally intractable [9–11]. In order to cope with the intractability of the problem, two main approaches were developed. First approach tries to solve the problem by developing different techniques for heuristic allocation of the continuous resource to feasible sequence of tasks [11–14]. Another one, takes advantage of the discretisation of continuous resource proposed in [9, 14]. The last approach was successfully used in [10, 15–22] for solving the DCSP and also within a simulated annealing algorithm for solving the discrete-continuous resource-constrained project scheduling problem in [9]. In the following subsection the discretisation of the continuous resource is discussed.

9.2.1 Discretisation of the Continuous Resource

As it follows from the definition of the DCSP, tasks require two types of renewable resources for their processing - discrete ones and an additional continuous resource. The assumption of continuity of the additional resource is fundamental for the DCSP. The continuous resource can be allocated to a task at time t in any amount from given interval. The number of such allocations for any DCSP is infinite. The idea of continuous resource discretisation is based on an assumption that the number of available allocations of the continuous resource to task J_i is finite and equals W_i. Such assumption results in the discretisation of the continuous resource and allows to treat it as an additional discrete resource. Let us denote the discretized allocation of continuous resource to task J_i at time t by $u_i^{l_i(t)} \in (0,1]$, $l_i(t) = 0, 1, 2, \ldots, W_i$. With the disposal of this allocation of the continuous resource task J_i will be processed with rate which can be calculated from equation:

$$\dot{x}_i(t) = f_i\left(u_i^{l_i(t)}\right) \tag{9.13}$$

It is assumed that if $l_i(t) = 0$, then zero amount of the continuous resource is allocated to task J_i, and therefore, task J_i is not processed. The completion of task J_i requires that $x_i(C_i) = \tilde{x}$. It is also assumed, without loss of generality, that $\sum_{i=1}^{n} u_i^{l_i(t)}(t) \leq 1$ for every $t > 0$. Although, it is allowed in general to change the allocation of continuous resource during the processing of task, however in practice, it is often assumed for simplicity that the allocation of additional resource remains constant during the whole processing of task. Such assumption is equivalent to the same assumption in $\mathbf{P_J}$. Assuming $u_i^{l_i(t)} = u_i^{l_i}$ (where t is any time moment during the processing of J_i), the time required to process task J_i using $u_i^{l_i}$ units of the additional resource is given by equation:

$$\tau_i^{l_i} = \frac{\tilde{x}_i}{f_i(u_i^{l_i})} \tag{9.14}$$

In such case, it can be assumed that each task is carried out in processing mode l_i which specifies the amount of the additional resource $u_i^{l_i}$, which will be allocated to the task, and therefore, the time in which the task will be processed. The amount $u_i^{l_i}$ can be defined in different ways, e.g. the simplest way would be to assume that:

$$u_i^{l_i} = l_i / W_i. \tag{9.15}$$

It is easy to see, that high discretisation level, i.e. larger values of W_i, would result in a greater similarity to the continuous resource. However, the large number of task processing modes would make the problem harder to deal with. Therefore, a discretisation level which would ensure the highest performance and accuracy of the algorithm using the discretised resource should be established.

As the result of the continuous resource discretisation, a special case of the Multi-Mode Resource-Constrained Project Scheduling Problem (MRCPSP) is obtained. The problem was investigated in numerous papers, e.g. [23–25] (for the survey see [26]), and proved to be NP-hard in [27].

Although the discretisation of continuous resource eliminates the need to solve the nonlinear programming problem, however, it does not lead to the problem which is equivalent to any of the considered DCSPs. It is obvious that the optimal allocation D_S^* determined for the optimal schedule (S^*, D_S^*) may not belong to the finite set of available allocations obtained after the discretisation of continuous resource. Also, the discretisation does not reduce the computational complexity of the primary DCSP, since, in general, scheduling independent tasks with fixed processing times on two identical parallel machines is NP-hard [10].

In the next subsection, the formulation of the DCSP with Continuous Resource Discretisation (DCSPwCRD) is given.

9.2.2 *Formulation of Discrete Continuous Scheduling Problem with Continuous Resource Discretisation (DCSPwCRD)*

The DCSPwCRD, denoted as Θ_Z for the purpose of further discussion, is formulated in the same way as in [10]. Namely, let $\boldsymbol{J} = \{J_1, J_2, \ldots, J_n\}$ be a set of nonpreemptable tasks, with no precedence relations and release dates $r_i = 0$, $i = 1, 2, \ldots, n$, and $\boldsymbol{P} = \{P_1, P_2, \ldots, P_m\}$ be a set of parallel and identical machines, and there is one additional renewable discrete resource in amount $U = 1$ available. A task J_i can be processed in one of the modes $l_i = 1, 2, \ldots, W_i$ (W_i – the number of processing modes of task J_i), for which J_i requires a machine from $\boldsymbol{P}$ and amount of the additional resource known in advance. The processing mode of J_i cannot change during the processing. For each task two vectors are defined: a processing times vector $\boldsymbol{\tau}_i = [\tau_i^1, \tau_i^2, \ldots, \tau_i^{W_i}]$, where $\tau_i^{l_i}$ is the processing time of task J_i in mode $l_i = 1, 2, \ldots, W_i$ and a vector of additional resource quantities allocated in each processing mode $\mathbf{u}_i = [u_i^1, u_i^2, \ldots, u_i^{W_i}]$. The total amount of the continuous resource used by tasks J_i at any time t within a schedule cannot exceed U.

The problem is to find the sequence of tasks from $\boldsymbol{J}$ on machines from $\boldsymbol{P}$ and task processing modes such that minimize given criterion of scheduling Q.

As it was mentioned in the previous subsection, the formulated problem is a special case of the more general MRCPSP, which is known to be NP-hard [27].

An instance I_{Θ_Z} of problem Θ_Z is obtained by specifying the values of all its parameters: n, m, W_i, vectors u_i and τ_i, $i = 1, 2, \ldots, n$, and criterion of optimality Q.

9.3 Heuristic Algorithms for Solving the DCSP

In [28], two heuristics: a Multi-Start Iterative Improvement Algorithm (MSIIA) and a Random Sampling Technique (RST) were proposed to solve the DCSP. A special case of the DCSP with concave processing rate functions $f_i = c_i u_i^{1/\alpha}$, $\alpha > 1$, $i = 1, 2, \ldots, n$ and an objective to minimize the makespan was considered. The results of computational experiment were compared with the results obtained by a Simulated Annealing algorithm (SA). In the experiments, $m \in \{2, 3, 10\}$, $n \in \{10, 20, 50, 100\}$ when $\alpha = 2$, and $m \in \{2, 3\}$, $n \in \{10, 20, 50\}$ when $\alpha = 3$. A brief description of the MSIIA and RST is given below.

In the MSIIA, a current solution S_a is replaced by the first solution S_b that improves S_a. If there is not such solution in the neighbourhood $N(S_a)$, i.e., if S_a is a local optimum, the MSIIA restarts at another initial solution. This procedure is repeated until the stopping condition (available computing time) is satisfied. The neighborhood generation mechanism and stop criterion are the same as in the SA.

In the RST, the next solution is randomly generated from the current one in order to find a solution with the smallest value of the cost function. The neighborhood generation mechanism and stop criterion are the same as in the SA.

As it was mentioned above both heuristics borrow neighborhood generation mechanism from the SA. This allows for easy comparison of their performance. Unfortunately, in almost all cases the SA performed better than the MSIIA and the RST. For smaller numbers of machines the RST gave better solutions than the MSIIA, however, for larger values of m it performed the worst.

In [1] two heuristics $H_1(h)$ and $H_2(A)$ for solving any DCSP with an objective to minimize the makespan were proposed. The description of the heuristics is given below.

In Algorithm $H_1(h)$, a preliminary allocation of continuous resource to the tasks is defined according to a chosen heuristic function h. Having the preliminary continuous resource allocation, the processing times of tasks are calculated, and a schedule of n nonpreemptable tasks on m parallel machines to minimize the makespan is determined. Since in general, the problem of scheduling n tasks on m parallel machines is NP-hard, the schedule can be found using either an optimal pseudo-polynomial dynamic programming algorithm proposed in [29], or one of the available heuristics. After the preliminary schedule has been obtained, an attempt to improve it is made by allocating the same amount of continuous resource u_j to all tasks, assigned to machine j, i.e. it is assumed that $u_i = u_j$ for $i \in J_j$, where J_j is the set of tasks assigned to machine j, $j = 1, 2, \ldots, m$. Additionally, it is assumed, that all machines finish at the same time. In order to determine a new schedule, new values of the continuous resource u_i to be allocated to the tasks should be calculated by solving the following system of equations:

$$\sum_{i \in J_j} \frac{\tilde{x}_i}{f_i(u_j)} = M, \quad j = 1, 2, \ldots, m, \tag{9.16}$$

$$\sum_{j=1}^{m} u_j = 1, \tag{9.17}$$

$$u_j \geq 0, \quad j = 1, 2, \ldots, m, \tag{9.18}$$

Having calculated new values for u_i and task processing times, new schedule is determined using equation:

$$\tau_i = \frac{\tilde{x}_i}{f_i(u_i)} \tag{9.19}$$

Because the same amount of continuous resource has been allocated to all tasks assigned to the same machine, the order of tasks on the machines in new schedule can be arbitrary.

Another heuristic proposed in [1] is denoted as $H_2(A)$. It uses the concept of feasible sequences and defines a very general approach towards discrete-continuous scheduling problems. Its effectiveness depends strongly on the quality of algorithm A which is designated to generate a set of feasible sequences S^F for a given problem instance I. As an example algorithm, which can be used for this purpose, Algorithm A_1 is proposed. The general idea of Algorithm A_1 is to create a set of feasible sequences S^F, using the property that the number of Z_k's, in which a task appears, grows with the processing demand of that task. After set S^F has been created, Algorithm $H_2(A)$ requires to find a semi-optimal schedule for each feasible sequence S $\in S^F$ by solving problem **P** A semi-optimal schedule of the minimum length is chosen as the solution to the problem.

Both described algorithms can find optimal schedules for some classes of problems. The first approach is less time consuming and explicitly uses results from the classical scheduling theory. The second one defines a general method for finding solutions to discrete-continuous scheduling problems.

In [30], a *Simple Heuristic* (SH) to solve a special case of the discrete-continuous scheduling problem with processing rate functions $f_i = u_i^{1/\alpha_i}$, $\alpha_i \in \{1,2\}$, $i = 1, 2, \ldots, n$ is introduced. SH exploits property **p-2** given in Sect. 8.8.3.2 while assigning tasks to machines. Recall, that property **p-2** states that in an optimal schedule at most one task from set A_1 ($\alpha_i = 1$) is processed at a time. The easiest way to fulfill this requirement is to assign all the tasks $J_i \in A_1$ to a single machine. On the other hand, the problem with $\alpha_i \in \{1, 2\}$ can be reduced to the problem with identical values of $\alpha_i^{'} = 2$ by calculating the values of processing demands of tasks using Eq. (8.19). When the identical values of α'_i are obtained, it is sufficient to allocate identical amounts of the continuous resource to each machine in order to obtain a semi-optimal schedule. Thus, in the first step of SH, the modified processing demands $\tilde{x}_i^{'}$ of tasks $J_i \in A_2$ ($\alpha_i = 2$) are calculated using Eq. (8.19) and demand $\tilde{x}_{n_2+1} = \sum_{i \in A_1} \tilde{x}_i$ (it is assumed that there is a task J_{n_2+1}, with processing demand $\tilde{x}_{n_2+1}$ equal the sum of processing demands of all tasks $J_i \in A_1$ with $\alpha_i = 1$). In the second step, the tasks are assigned to the machines according to non-increasing values of $\tilde{x}'_i$. In the third step, the value of makespan is calculated using Eq. (8.18). SH yields slightly worse results than SA (Simulated Annealing), TS (Tabu Search), and GA (Genetic Algorithm) designed for solving the same problem. The description of mentioned metaheuristics can be found in Sect. 9.4. Because the complexity of SH is $O(nlogn)$, its computational effort is at least 2×10^3 times smaller than for any of mentioned metaheuristics. This makes SH a reasonable alternative for the metaheuristics in situations, when a suboptimal solution is required in a relatively short time.

In [13], a heuristic procedure HUDD was used in the TS for continuous resource allocation. The idea of the HUDD heuristic is based on a uniform distribution of the processing demand of each task respectively to the number of combinations Z_k in which this task occurs. For this purpose, the number q_i of all combinations Z_k containing task J_i is counted, and the processing demand $\tilde{x}_i$ of J_i is divided by q_i (i.e. $\tilde{x}_{ik} = \tilde{x}_i / q_i$) in order to distribute it uniformly over all those parts of the task. Having obtained the parts of the processing demands of the tasks $\tilde{x}_{ik}$ in each combination Z_k, $k = 1, 2, \ldots, n - m + 1$, the length M_k of the k-th interval is then calculated as the unique positive root of equation:

$$\sum_{i \in Z_k} f_i^{-1}(\tilde{x}_{ik} / M_k) = 1 \tag{9.20}$$

where $\tilde{x}_{ik}$ is the part of the processing demand of task J_i, $i = 1, 2, \ldots, n$, executed in the considered interval k. Finally, the makespan M of a schedule is calculated from equation:

$$M = \sum_{k=1}^{n-m+1} M_k. \tag{9.21}$$

It should be mentioned, that each function f_i must be a bijection in order to compute the reverse function. Moreover, even then, solving Eq. (9.20) may not be quite simple, however in many cases it can be solved analytically. In [12], four heuristics were compared on a basis of an extensive computational experiment, and the HUDD turned out to be the most efficient for the considered problem. The HUDD was used in the TS as the procedure for the continuous resource allocation and discussed in [13]. Although, the combined heuristic denoted as TS-HUDD was unable to find the optimal makespan, however the average relative deviation from optimal solution was about 1%–1,5% and the maximal one—about 4%–4,5%. The computational time needed for TS-HUDD to yield a solution was tens of thousands of times smaller than the time needed for TS-OPT which used a specialized solver CFSQP 2.3 [31] for finding the optimal allocation of the continuous resource. Moreover, the computational time ratio between the two algorithms grew with the problem size, which can be counted among the advantages of the HUDD.

In [14], the HUDD, denoted as HUDD-PS, was used for solving the general case of the discrete-continuous resource-constrained project scheduling problem (DCRCPSP). In this work, the quality of the solutions found by HUDD-PS was compared to the optimal solutions determined by improved version of a solver CFSQP 2.5 [32]. The results show that the proposed HUDD-PS heuristic can be considered as quite effective for the analyzed problem. The average relative deviation from optimum did not exceed 6%, whereas the maximal relative deviation oscillates around 12%. Moreover, these values did not increase with the growth of the problem size, which might suggest that the heuristic was independent of the number of activities, at least for the problem sizes considered. Thus, taking into account the acceptable quality of the solutions found and significantly smaller computing time (from about 1000 times for 10 activities up to about 10000 times for 20 activities), the HUDD may be successfully used for solving the continuous part of the problem.

9.4 Metaheuristics for Solving the DCSP

9.4.1 TS, SA, and GA as Local Search Metaheuristics for Discrete-Continuous Scheduling Problems

As it is known from Sect. 8.7, the general methodology for solving any DCSP assumes firstly, determining feasible sequences of tasks on machines, and secondly, determining optimal allocation of continuous resource to these sequences. The second stage of solving the DCSP is formulated as a convex mathematical programming problem with linear constraints which can be solved using the appropriate solver. Because, in general, the cardinality of the POS grows exponentially with the number of tasks (see the definition of POS in Sect. 8.7), three metaheuristics (SA—simulated annealing, TS—tabu search, and GA—genetic algorithm) described in [30, 33] were proposed to solve the problem with an objective of the makespan minimization. In [33], the problem of continuous resource allocation was solved using specially adopted solver, whereas in [30]—analytically, using proved in this paper properties of optimal schedules. In both papers, concave processing rate functions were considered, and the search space of all metaheuristics was limited to POS. It has been proved in [1], that for concave functions f_i, $i = 1, 2, \ldots, n$, a set containing all feasible sequences composed of $p = n - m + 1$ m-element combinations of tasks comprises the POS.

In the experimental part of both works, a special case of the discrete-continuous scheduling problem with processing rates functions of the form $f_i = u_i^{1/\alpha_i}$, $\alpha_i \in \{1, 2\}$, $i = 1, 2, \ldots, n$ and an objective to minimize the makespan was solved. Only the case of $n_1 > 0$ and $n_2 \geq m$ was considered (out of four possible cases discussed in Sect. 8.8.3.2), where m is the number of machines, and n_1 and n_2 are the numbers of tasks with processing rate functions $f_i = u_i$, i.e. $\alpha_i = 1$, and $f_i = u_i^{1/2}$, i.e. with $\alpha_i = 2$ respectively. Processing demands of the tasks have been generated from the interval [1, 100] with a uniform distribution. Values of α_i from the set {1, 2} were generated randomly with equal probability. In order to ensure similar computational effort performed by each heuristic, the stop criterion was defined as the number of solutions visited and set for 2000.

In [33], the results of numerous computational experiments, performed to examine the application of TS, SA and GA to the general discrete-continuous scheduling problem are presented. The aim of the experiments was to analyze the convergence and to set the parameters of the metaheuristics. Conclusions concerning the best neighbourhood generation mechanism in SA, tabu list management in TS, recombination operators in GA and parameter settings for all algorithms as well as their convergence behavior were provided. The results of the experiments showed that TS was more efficient than SA and GA. TS found the largest number of optimal solutions and showed smallest deviation from optimum for all sizes of the problem. As it was mentioned above, the allocation of continuous resource to each solution visited in the solution space was determined by solving a convex mathematical programming problem. For this purpose, a specially adopted solver CFSQP 2.3 (A C Code for Solving (Large Scale) Constrained Nonlinear (Minimax) Optimization Problems, Generating Iterates Satisfying All Inequality Constraints) [31] was applied. The solver stopped when the absolute difference in consecutive values of the objective function was less than or equal to 10^{-5}. The average time of processing a single instance of the size $n = 15$, $m = 2$ was approximately 5000 s. The main reason for such large processing time needed to yield a solution was the computational effort required by the solver to find a solution to the convex mathematical programming problem.

In [30], all considered metaheuristics used the specific properties of optimal schedules for the considered problem, proved in the discussed paper and recalled in this work in Sect. 8.8.3.2. Based on property **p-2** (see Sect. 8.8.3.2), the search space of the metaheuristics was limited only to feasible solutions in which at most one task with $\alpha_i = 1$ is scheduled at a time. A specially designed Solution Feasibility Test (SFT), based on property **p-2**, was incorporated into each metaheuristic in order to reduce the search space by rejecting solutions which did not comply with it. The optimal allocation of continuous resource to the tasks in a given feasible sequence was determined analytically using Eq. (8.18). The efficiency of heuristics under consideration was compared using a relative deviation of a solution found by metaheuristic from an optimal solution which was determined as follows. Firstly, all possible assignments of tasks to machines were generated. Next, for a given assignment, permutations of tasks on each machine were obtained. A permutation (solution) which passed the SFT was identified as an optimal schedule for this assignment. The best among all solutions which passed the SFT was considered to be the optimal solution. The computational experiments showed that owing to the reduction of the search space due to the SFT and possibility of solving the continuous resource allocation analytically all metaheuristics performed much more efficiently compared to the case described in [33], where the specialized solver was used to determine the continuous resource allocation. The average time of processing a single instance of the size $n = 15$, $m = 2$ was reduced from 5000 s, as it was when the solver was used, to about 1,58 s. It took on average 0,95 s, 0,75 s, and 3,04 s for SA, TS, and GA respectively to find a solution. In the experiments, the best results were obtained by GA, however, both average and maximum relative deviations from optimum were small for SA and TS as well. The results show that all metaheuristics under consideration performed better than the Multi-Start Iterative Improvement Algorithm (MSIIA) and Random Sampling Technique (RST) proposed in [28], as well as the Simple Heuristic (SH) proposed in [30], which are briefly described in Sect. 9.1.

In [34], the same three metaheuristics were used to solve the same special case of the DCSP, i.e. the case with task processing rates of the form $f_i = u_i^{1/\alpha_i}$, $\alpha_i \in \{1, 2\}$, $i = 1, 2, \ldots, n$, however, with an objective to minimize the mean flow time (the respective problem is discussed in Sect. 8.10). For this problem, a feasible solution was represented by a feasible sequence of n combinations of tasks, such that first $n - m + 1$ combinations contained exactly m elements and the consecutive combinations $n - m + 2$, $n - m + 3$, ..., $n - 1$, n contained $m - 1$, $m - 2$, ..., 1 element respectively. The same specialized solver CFSQP 2.3 was used to find the continuous resource allocation for each solution visited in the solution space. The results of computational experiments showed, that TS performed best, finding the largest number of best solutions and showing smallest deviation from optimum for all the problem sizes. The advantage of the tabu search algorithm increases with the growth of the number of tasks, and reaches 100% for 20 tasks (i.e. for each instance of the problem, tabu search finds the best solution). However, the average and maximum relative deviation of the solutions found by GA from the best solution were the smallest.

In [35], the same three metaheuristics under consideration were also used to solve the maximum lateness minimization problem. For this problem, as well as for the mean flow time minimization problem, a feasible solution was represented by a feasible sequence of n combinations of tasks, such that first $n - m + 1$ combinations contained exactly m elements and the consecutive combinations $n - m + 2$, $n - m + 3$, ..., $n - 1$, n contained $m - 1$, $m - 2$, ..., 1 element respectively. Only limited experiments were carried out because of the time consuming continuous part of the problem. The solution of the convex mathematical programming problem needed for finding an optimal continuous resource allocation was too time-consuming in this case. It took the TS about 24 h to solve one instance of the problem with $n = 10$ tasks, when the number of solutions visited was limited to 1000. For this reason, the experiment was carried out only for $n = 10$ tasks scheduled on $m \in \{2, 3, 4\}$ machines, whilst the number of generated instances for each combination $n \times m$ was 21, 14, and 7, respectively. The results showed, that tabu search found solutions with the best maximum lateness most often. However, for 2 and 3 machines, the GA achieved better maximal deviation.

In [9], an original heuristic approach to the problem of the continuous resource allocation was proposed. The approach exploits the idea of discretisation of the continuous resource described in Sect. 9.2.1. Now, the discrete resource is allocated in amounts known in advance, i.e. a task is performed in some mode, characterized by the amount of the resource and task processing time which correspond to the chosen mode. The amounts of the resource should be pre-calculated using some method for the discretisation, e.g. using Eq. (9.15), which was also used in the discussed research. Recall from Sect. 9.2.1, that as the result of the continuous resource discretisation, a special case of the Multi-Mode Resource-Constrained Project Scheduling Problem (MRCPSP) is obtained. Thus, having the continuous resource discretised, the problem is to find the sequence of tasks on machines and task processing modes such that minimize given criterion of scheduling. The proposed approach was used in the SA for solving the continuous resource allocation problem. Although, the SA was used to solve the discrete-continuous resource-constrained project scheduling problem (DCRCPSP) to minimize the makespan, nevertheless, we will recall the results of application of the discretisation of the continuous resource, since this approach can also be applied to the DCSP. In the experiment, a special case of the DCRCPSP was considered, where only one discrete resource R_1 was given. The resource requests of all activities were identical and equal 1, and there were no precedence constraints imposed. In fact, this special case reduces to a discrete-continuous machine scheduling problem, where the discrete resource is a set of parallel identical machines, with the number of machines equal to the number of available units of single discrete resource R_1. A version of the SA (SAM+), where the idea of the continuous resource discretisation was implemented, was compared with the version of SA (SAA) in which the continuous resource was allocated optimally with aid of the CFSQP 2.3 solver. In SAM+, a preliminary schedule, obtained for MRCPSP, was then improved by constructing a feasible sequence S^z and solving problem $\Omega(C_{\max})$ (see Sect. 8.8.3) for this sequence. The obtained schedule was the solution to the corresponding DCRCPSP. The experiment was performed for $n = 10$ activities, and one discrete resource available in various numbers of units $R_1 \in \{2, 5, 10, 15\}$. Processing rate functions of activities were also of the form $f_i = u_i^{1/\alpha_i}$, $\alpha_i \in \{1, 2\}$, $i = 1, 2, \ldots, n$. For simplicity, the same number of modes following from the continuous resource discretisation was assumed for each activity of the project, i.e. $W_i = W$, $i = 1, 2, \ldots, n$. Different values of parameter W were examined, from $W = 2$ up to $W = 100$. For each combination of parameters R_1 and W, 30 instances were randomly generated. The number of visited solutions, defining the stop criterion for both the algorithms, was set at 1000. The experiment showed that the distance of the results produced by SAM+ from the ones obtained by SAA was 4,2% on average, and did not exceed 5,5% for all problem sizes. For a few instances the SAM+ procedure was also able to find solutions of the same quality as SAA (up to 5 out of 30). However, the average computational time for one instance was about 190 s for SAM+, whereas for SAA it was about 13000 s on average. Thus, the main result of using the continuous resource discretisation approach is that it allows to reduce the computational time (up to 70 times compared to SAA) while losing not more than 5,6% on the quality of solutions. Moreover, it was made an observation that, the quality of results yielded by the SAM+ results was not influenced neither by the number of available units of the discrete resource R_1, or the discretisation level W. An increase of the number of modes W not always improved the quality of solutions, despite the fact that the allocations of the discretised resource approached in accuracy to the allocations of the continuous resource. It was established, that quite good results were obtained for $W = 5$ for all data sets considered in the experiment. The main conclusion was that the obtained results justified the use of the discussed discretisation approach for coping with the continuous part of the DCSP.

The successful application of the TS for solving the DCSP motivated researchers for its further development. In [36, 37], the TS was used for solving the DCSP with the makespan, the mean flow time, and the maximum lateness as the scheduling criteria. Three tabu list management methods: the Tabu Navigation Method (TNM) [38], the Cancellation Sequence Method (CSM) [39], and the Reverse Elimination Methods (REM) [39] were implemented and compared (a review of basic concepts and developments of TS one might find in [40]). In the tests, the TS again outperformed the SA and the GA. For the makespan minimization problem the results yielded by the TS were very close to optimum. In the experiments on tabu list management methods, TNM performed slightly better, than two other methods, which makes the TS very effective, regardless of the list management method applied.

In [41], further research was carried out to examine the performance of the TNM, CSM, and REM for the case of the makespan, mean flow time, and maximum lateness minimization. The performance of the TS combined with three beforementioned tabu list management methods was again compared to the performance of the SA and the GA. The experiment was carried out for task processing rate functions of the form $f_i = u_i^{1/\alpha_i}$, $\alpha_i \in \{1, 2\}$, $i = 1, 2, \ldots, n$, where the values of α_i were generated randomly with equal probabilities. The allocation of the continuous resource to the sequence of tasks was determined using the specialize solver CFSQP 2.3, which stopped when the absolute difference in consecutive values of the objective function was less than or equal to 10^{-3}. The stop criterion for the TS was defined as the number of solutions visited and set for 1000. In the case of the makespan, the results were compared to optimal solutions (found by the full enumeration procedure), in the cases of the mean flow time and the maximum lateness, the results were compared to the best solutions found by the metaheuristics tested. Because of the hardness of the problem, it took about 0,6 s for the solver to find a continuous resource allocation to 10 tasks and about 8 s to 20 tasks sequenced on 2 machines. This also results in almost identical computational times required by all metaheuristics tested since the majority of their computational effort is devoted to the calculation of the objective function. For each problem size tested, TS found an optimal solution in about 50% of instances, The average relative deviation from optimum did not exceed 0,1% in any case, and the maximum relative deviation oscillates around 1%. The results obtained by TS were significantly better than the ones obtained by the two other metaheuristics, both in terms of the number of optimal solutions found, and in terms of the average relative deviation. It was observed in the experiment, that the growth of the number of tasks did not result in a significant deterioration in the average relative deviation, which was not the case for maximal relative deviation. On the other hand, the growth of the number of machines, under a fixed number of tasks, had greater influence on the results. The results of the experiments show that the tabu navigation method performed best for the considered problem. The number of optimal solutions found by this method was the greatest for each problem size, whereas the average relative deviation from optimum was the smallest. The maximum relative deviation from optimum for the TNM was smaller or at most equal to the deviations obtained for the two other methods. Generally, the results obtained by all the three methods were of high quality. Using all three tabu list management methods, the TS was able to find 50–70 out of 100 optimal solutions for each problem size, and the average relative deviation of the solutions found was below 0.1%. Such results prove that TS is very effective for the considered class of problems. Finally, although the TNM performs best for the considered problem, the results obtained for all the three methods are, in general, of a similar quality. Thus based on their observations, the authors conclude, that other aspects of the tabu search strategy (the solution representation, the neighborhood generation mechanism, and the starting solution) may have even stronger impact on the effectiveness of this metaheuristic than the tabu list management method itself.

In [13], a heuristic procedure—HUDD was used by the TS for the continuous resource allocation (the description of the HUDD procedure is given in Sect. 9.1). Although, the combined heuristic denoted as TS-HUDD was unable to find the optimal makespan, however the average relative deviation from optimal solution was about 1%–1,5% and the maximal one—about 4%–4,5%. The computational time needed for TS-HUDD to yield a solution was tens of thousands of times smaller than the time needed for TS-OPT which used a specialized solver CFSQP 2.3 [31] for finding an optimal allocation of the continuous resource. Moreover, the computational time ratio between the TS-HUDD and TS-OPT grew with the problem size. Such independence of the performance of TS-HUDD from the size of the problem can be considered as the advantages of the proposed heuristic approach.

9.5 Minimization of the Resource Usage in the DCSP

Originally, the main goal of the discrete-continuous scheduling was the minimization of criteria related to the time necessary to carry out a set of tasks. These criteria were: the makespan, the mean flow time, and the maximum lateness, which have been already discussed in Sects. 8.8–8.10. It turned out, that the methodology developed for the DCSP can be successfully applied to solve problems with other criteria. Recently, the results of the research on the minimization of the use of the continuous resource were published. It was pointed out in [5] that reducing the usage of the additional resource could contribute to the minimization of the costs arising from the use of this resource. An example of the resource minimization could be reducing all kinds of power required to drive the machines, also the amount of gas used in steel mills to heat the metal, the amount of common memory used in multiprocessor computer systems [1, 42], or the amount of energy for variable voltage processors used in portable electronic devices [45].

In [5], a research on the continuous resource level minimization in the discrete–continuous scheduling was made. Problem $\Omega(U_{\max})$, which minimizes the usage of the continuous resource required to process the tasks within a given upper bound on makespan $\widehat{C}$, was formulated. Based on Riemann's integration theory [43], a theorem stating that in an optimal solution the total amount of the continuous resource in use in any moment of time t is constant within a schedule. There was also proved another theorem on the relation between $\Omega(U_{\max})$ and $\Omega(C_{\max})$, stating that if in a solution the upper bound on makespan $\widehat{C}$ and optimal makespan $C^*_{\max}$ are equal in value, then the upper bound on the use of the continuous resource $\widehat{U}$ is optimal, i.e. if $\widehat{C} = C^*_{\max}$, then $\widehat{U} = U^*_{\max}$ and vice versa. These theorems helped to prove some other properties of problem $\Omega(U_{\max})$ and design an approximation algorithm which yields a solution with any desired precision. There were also proved properties which specify how to process the tasks on the machines depending on the tasks' processing rate. Namely, for the case of convex processing rate functions, the usage of the continuous resource is minimized when all tasks are processed on the same machine in an arbitrary order (identical property concerns the minimization of makespan $C_{\max}$ and the total flow time F, see Sect. 8.8.1). In this case, the optimal amount of the continuous resource $U^*_{\max}$ required to process all tasks within given upper bound on schedule $\widehat{C}$ can be determined according to the equations below:

$$f_i^{-1}\left(\frac{\tilde{x}_i}{\tau_i}\right) = f_{i+1}^{-1}\left(\frac{\tilde{x}_{i+1}}{\tau_{i+1}}\right), \quad i = 1, 2, \ldots, n-1, \tag{9.22}$$

$$\sum_{i=1}^{n} \tau_i = \widehat{C}, \tag{9.23}$$

$$U^*_{\max} = f_i^{-1}\left(\frac{\tilde{x}_i}{\tau_i^*}\right), \quad i = 1, 2, \ldots, n, \tag{9.24}$$

where τ_i^* is the processing time of task J_i in the optimal schedule. In the case of concave functions and $n \leq m$, the usage of the continuous resource is minimized by processing each task on another machine (identical property concerns the minimization of makespan $C_{\max}$, see Sect. 8.8.2), and allocating the continuous resource in amounts given by the following equations:

$$u_i^* = f_i^{-1}\left(\frac{\tilde{x}_i}{\widehat{C}}\right), \quad i = 1, 2, \ldots, n, \tag{9.25}$$

$$U^*_{max} = \sum_{i=1}^{n} u^*_i. \tag{9.26}$$

Based on the proved properties of problem $\Omega(U_{max})$, for the case $n \geq m$ the set of all feasible sequences S^z comprised by all sequences containing $p \leq n - m + 1$ combinations Z_k was reduced to the set of all feasible sequences $\boldsymbol{S}^z$ which contains $p = n - m + 1$ combinations Z_k. Such property of $\boldsymbol{S}^z$ complies with the definition of the POS of $\Omega(C_{max})$ given in Sect. 8.8.3. It was also proved that the cardinality of $\boldsymbol{S}^z$ for the case $n \geq m$ can be determined using the equation below:

$$|S^Z| = \frac{n! \cdot m^{n-m}}{m!}. \tag{9.27}$$

In the paper, there were given two additional formulations of problem $\Omega(U_{max})$, denoted as $\Omega'(U_{max})$ and $\Omega''(U_{max})$, which are more suitable for handling by the Sequential Quadratic Programming (SQP) method [6] provided by MATLAB Optimization Toolbox. The formulation of problem $\Omega'(U_{max})$ was obtained through standard transformation of $\Omega(U_{max})$ into a problem with the simpler to optimize linear objective function. Problem $\Omega''(U_{max})$ was formulated in order to facilitate processing of some input data. It was obtained from $\Omega(U_{max})$ by adding a constraint which explicitly imposes constant level of the continuous resource used within a schedule (the property of optimal schedule) and replacing the equation for the primary objective function by the equation for the total amount of the continuous resource assigned to any, e.g. first, interval of the schedule. In the paper, there was proposed an algorithm which finds an optimal resource allocation for given sequence S^z on the way of solving one of problems $\Omega(U_{max})$, $\Omega'(U_{max})$ or $\Omega''(U_{max})$. In practice, standard optimization procedures are often used for solving nonlinear problems of mathematical programming. Such standard optimization procedures had been used in an approximation algorithm, additionally proposed in the paper, which solves problem $\Omega(U_{max})$ with an arbitrarily small error $\varepsilon(M_{min})$, i.e. $U_{max} - U^*_{max} < \varepsilon(M_{min})$, where M_{min} is a minimal length interval equal to the smallest number which can be handled by a computer at hand. The proposed approximation algorithm generates all feasible sequences from $\boldsymbol{S}^z$ and for each sequence determines the allocation of the continuous resource with an absolute approximation error $\varepsilon(M_{min})$. Because, the cardinality of set $\boldsymbol{S}^z$ grows exponentially with the size of the problem, a fast heuristic algorithm AS(π_1, π_2) for solving $\Omega(U_{max})$ was proposed. In the heuristic, the procedure for generating all sequences $S^z \in \boldsymbol{S}^z$ was replaced by a simpler one which composes sequence $S^z \in \boldsymbol{S}^z$ using two sequences of tasks π_1 and π_2 in which tasks from $\boldsymbol{J}$ are placed in an arbitrary order. Sequence π_1 determines the order in which tasks start their processing in a schedule, and π_2 determines the order in which they finish their processing. For the sequences $S^z \in \boldsymbol{S}^z$ obtained in this way, problem $\Omega(U_{max})$ is solved. The performance of the proposed methods and algorithms was examined in the experiment, where problems $\Omega(U_{max})$, $\Omega'(U_{max})$ and $\Omega''(U_{max})$ with convex and concave processing rate functions were solved. In the case of concave functions, the SQP method was used for solving $\Omega(U_{max})$, $\Omega'(U_{max})$ and $\Omega''(U_{max})$ problems in order to determine which formulation of the problem results in better performance. As an input data, concave functions of the form $f_i(u_i) = c_i \cdot u_i^{1/\alpha_i}$, where $c_i > 0$ is a positive real number, three sets of α_i: $\alpha_i \in \{1, 2\}$, $\alpha_i \in \{1, 2, 3\}$, $\alpha_i \in \{1, 2, 3, 4\}$, the number of machines $m \in \{2, 3, 4\}$, and the number of tasks $n \in \{10, 15, 20\}$ were used to generate 1800 instances of the problem. The values of processing demands of tasks $\tilde{x}_i$ and coefficients c_i, $i = 1, 2, \ldots, n$ were generated at random with a uniform distribution from an interval between 0 and 1. The upper bound on the schedule length $\widehat{C}$ was set at 10. The value of minimal length interval M_{min} was set at 10^{-6}, which ensured relative approximation error of the solutions not greater than 10^{-6} %. All the experiments were performed on a PC computer with AMD Athlon 2000+ processor and 1 GB RAM memory in the MATLAB R14 environment. The quality of solutions found by the tested algorithms was evaluated using relative error, calculated as $((U_{max} - U'_{max})/U'_{max}) \cdot 100\%$, where U_{max} is the resource-amount determined by the SQP method for a given instance, and U'_{max} is the best among all values of U_{max} obtained for the same instance. The mean relative errors of solutions found for problem formulations $\Omega(U_{max})$, $\Omega'(U_{max})$ and $\Omega''(U_{max})$ were similar and were in the range between 0,01% and 0,88% (when $\alpha_i \in \{1, 2\}$ and $\alpha_i \in \{1, 2, 3\}$), and between 0,01% and 2,89% (when $\alpha_i \in \{1, 2, 3, 4\}$). However, the time required to yield a solution in case of $\Omega(U_{max})$ (5 s–41,3 s) was 3–4 times greater than in the case of $\Omega'(U_{max})$ and $\Omega''(U_{max})$ (1,5 s–12,9 s). In the case of heuristic AS(π_1, π_2), the mean relative errors were in the range between 28% and 57,7%. And the relative error was calculated as $((U_{max} - U'_{max})/U'_{max}) \cdot 100\%$, where U_{max} is the resource-amount determined by AS(π_1, π_2) for a given pair (π_1, π_2), and U'_{max} is the best among values of U_{max} obtained for all pairs (π_1, π_2). Unfortunately, the performance time of AS(π_1, π_2) has not been provided in the paper. There were also conducted experiments for convex processing rate functions, although these functions have only limited theoretical value. In the case of convex functions, there does not exist the problem of sequencing the tasks on machines, since an optimal schedule is obtained when all tasks are processed on the same machine in an arbitrary order, and the amounts of the continuous resource to be allocated to the tasks can be determined by solving Eqs. (9.22)–(9.24). Three methods for solving the problem of continuous resource allocation were compared: an original approximation method proposed by the authors, a MATLAB approximation method, and a MATLAB symbolic method. As an input data, convex functions of the form $f_i(u_i) = c_i \cdot u_i^{1/\alpha_i}$, where $c_i > 0$ is a positive real number, three sets of α_i: $\alpha_i \in \{1, 2\}$, $\alpha_i \in \{1, 2, 3\}$, $\alpha_i \in \{1, 2, 3, 4\}$, the number of machines $m \in \{2, 3, 4\}$, and the number of tasks $n \in \{5, 10, 20, 30, 100, 10^5\}$ were used to generate 1800 instances of the problem. The mean time required to solve Eqs. (9.22)–(9.24) for most of the problem sizes by the proposed approximation method was less than 0,1 s. The mean time required by MATLAB approximation methods in most cases was close to 0,1 s. Unfortunately, the MATLAB symbolic method was useless in the cases with $\alpha_i > 2$ and the number of tasks $n > 5$, because of the excessive computational time.

In [44], the authors show how the methodology developed for the discrete-continuous scheduling can be applied to manage power consumption in computer systems. The power consumption issue is an important performance factor in all kinds of battery-powered multiprocessor devices where multiple independent tasks are being simultaneously carried out. The efficient use of energy can not only extend the operating time of such devices, but also prevent them from overheating. In the paper, a commonly used task processing model, describing the relation between the speed of processing a task and the power consumed during this processing, is generalized to the model where the rate of processing may change in time and depends on the temporal power allocation. One of the advantages of the proposed model is the possibility of describing much broader class of power usage functions, whereas in commonly used model, convex functions of power $\alpha > 1$ are mainly assumed (for CMOS technology $\alpha = 3$). Another advantage is the possibility to describe tasks with various power usage functions, which is useful when the power consumption depends on the type of microprocessor instruction being executed. On the contrary, in the previous model it was assumed the same power usage functions for all tasks. The proposed model of task processing was used to formulate a problem corresponding to practical situations where power consumption minimization is in focus. There were also discussed some theoretical results for the case with a single constraint on the energy consumption at every time moment t, i.e. the constraint on power consumption, and the case with doubly constrained energy consumption, where instantaneous and the total power consumption were constrained. Perceiving the energy as scarce doubly constrained resource is very important in power management, because it allows for better modelling of the real practical situations. The paper discusses some properties of time-optimal schedules and suggests the ways of obtaining optimal or suboptimal solutions to power management problem.

In [45], a problem of scheduling a set of preemptable tasks on a single variable voltage processor to minimize overall energy consumption was considered. This problem arises in autonomous devices driven by an independent energy source, e.g. portable electronic devices. When the capacity of the energy source is limited, it is crucial to carry out scheduled tasks in an expected time while consuming as little energy as possible. This type of problem is known in the literature as min-energy dynamic voltage scaling scheduling, or simply min-energy DVS scheduling [46–48]. In the problem, each task is characterized by its ready time, deadline, and the number of CPU cycles required to perform a task. A feasible schedule defines a task and its processing rate at time t within the schedule. The energy required to execute the task is assumed to be a convex function of the processing rate (at least of the second power). Two variants of the problem were considered: continuous and discrete. In the continuous variant, it assumed that the processing rate function is piecewise continuous with finitely many discontinuities and takes real values, whereas in the discrete one, it takes values from the set of the finite number of predefined processing rates. Therefore, the goal in the discrete variant is to determine a min-energy schedule using only these predefined rates. The contribution of the research is an optimal algorithm which had improved over the best previous one of time complexity $O(n^3)$ to $O(n^2 \log n)$. The algorithm constructs the optimal continuous schedule by successive approximation, based on an efficient partitioning of the set of tasks into subsets of high and low processing rates tasks with respect to some processing rate threshold, without computing the exact processing rate.

9.6 The Special Case of the DCSP

In [49], a continuous resource sharing (CRSHARING) problem was introduced. In practice, an example of continuous resource can be a limited computer memory or system's bandwidth which limits the distribution of data, necessary for carrying out computations, among processors. The problem concerns the allocation of continuous resource to m identical constant speed processors which have to process n tasks. The assignment of tasks to processors and the order, in which the tasks have to be processed, are given in advance. A task is processed in one or a few discrete time steps. During any time step, only one task can be processed by a processor. A task is processed at full speed, if it receives an appropriate amount of the continuous resource, called resource requirement $\hat{u}_i(t)$. Resource requirement $\hat{u}_i(t)$ determines the amount of the resource required to process one unit of the task's processing demand in one time step. The processing rate of a task depends linearly on the amount of the continuous resource it has been allocated in a time step, i.e. if a task receives an x-th fraction of whole amount of its resource requirement, where $x \in [0, 1]$, then it can be processed at the x-th fraction of full speed in that step. The total amount of the continuous resource allocated to processors at any time step t is limited and cannot exceed a given value, i.e.

$$\sum_{j=1}^{m} u_j(t) \leq 1 \tag{9.28}$$

where $u_j(t) \in [0, 1]$ is the amount of the continuous resource allocated to processor P_j at time step t. The problem is to determine the allocation of the resource to the processors in every time step which minimizes the makespan of the schedule subject to constraints on the resource (9.28). The considered problem is similar to the DCSP, because task processing rate can be described by a concave (linear) function f_i of the form:

$$f_i(t) = \min\left(\frac{u_j(t)}{\hat{u}_i(t)}, 1\right), \tag{9.29}$$

where $\hat{u}_i(t) \in [0, 1]$ is the resource requirement of task J_i. However, the considered problem is different from the DCSP, because the assignment of tasks to processors and the order in which the tasks are to be processed are imposed in advance. Although, it was assumed in the research, that tasks were of a unit size, however it was proved that the problem is NP-hard in the number of processors. Nevertheless, an optimal algorithm for solving the problem in $O(n^2)$ time for the case with $m = 2$ and assuming unit size tasks was proposed. For the case with the number of processors $m \geq 3$ and unit size tasks, an approximation algorithm with a worst-case approximation ratio of $2 - 1/m$ was also proposed.

9.7 Research on the Island Model of Computing

The performance of evolutionary algorithm (EA) to a great extent depends on its ability to cope with the obstacles that prevent or hinder the progress of the search. One of these obstacles is premature convergence, which results in getting stuck in a local optimum. Another one—too large search area, which is a natural obstacle on the way to global optimum. One of the ways to cope with these difficulties is to provide the appropriate level of population diversity. In the literature, we find a variety of approaches and techniques to maintain an appropriate level of the population diversification. For a comparative review of approaches to prevent premature convergence in GAs see [50]. One of the simplest ways to diversify the population, is to determine such size of the evolving population that ensures the highest efficiency of EA. Alternatively, one can initiate and perform search in various distant and separated from each other regions of the search area. In such case, the algorithm operates on multiple and independent to a certain degree sub-populations, which provide desirable diversity, and, in consequence, may contribute to faster convergence towards the global optimum. Such method of performing the search was the subject of interest of many researchers, e.g. [18–20, 51–54]. It is also known in the literature as an island model and is often reported as more effective, than a search performed on a single population, e.g. [55, 56]. The idea of increasing the population diversity by structuring it as autonomous and interacting with each other sub-populations was borrowed from the mathematical theory of population genetics, where an isolation by distance theory, pioneered in [57, 58], gave rise to different models of population structure. In an island model, proposed in [57], the whole population is divided into separated subpopulations (islands). All subpopulations are panmictic (i.e. all individuals are potential partners), and random migration of individuals among subpopulations is assumed. In evolutionary computation, the island model is viewed as a set of homogeneous or heterogeneous algorithms which autonomously evolve assigned to them subpopulations. Such islands are interconnected with each other according to some topological scheme (a ring, a torus, a hypercube etc.) which determines the migration of individuals among them. The migration among the islands is performed by some chosen *migration policy* which determines *migration interval* (the number of generations or fitness function evaluations after which the migration is carried out) and *migration size* (the number of individuals which is to be sent between sender and receiver islands). The described island model is well suited for processing in parallel, distributed, and agent systems which may provide greater effectiveness of search and reduction of the response time. The remainder of this chapter provides an overview of some studies on the effectiveness of evolutionary and genetic algorithms which exploit the generally understood island model, i.e. the model in which separated and interacting with each other subpopulations perform an evolutionary search.

In [52], the performance of distributed (DGA) and canonical (CGA) GAs was tested on the Royal Road functions (*R*1-4). In the DGA, originally proposed in [59], the subpopulations regularly exchange some of their solutions in regular intervals. In [59], two parameters of such migration of solutions were introduced: *migration interval* (*i*)—the number of generations between each migration, and *migration rate* (*r*)—the percentage of individuals selected for migration. In the experiments conducted in [52], the total size of population was set at 480, the number of subpopulations—at 24, and the size of each subpopulation—at 20. During each run of the DGA, 500 generations were evolved, which corresponds to 240480 of the fitness function evaluations (including the initial generation 0). Migration interval $i \in \{5, 10, 20, 50, 100, 500\}$, and migration rate $r \in \{0{,}1, 0{,}2, 0{,}5\}$, i.e. the subpopulations exchanged among themselves 10%, 20%, and 50% of their individuals during the migration phase. The case with $i = 500$ corresponded to the DGA without migration. In [52], the migration in the DGA was carried out between randomly chosen pairs of subpopulations, which was different to the hypercube topology, originally implemented in [59]. During the migration a fixed number $n_{mig} = r \cdot n$ (where n is the size of the subpopulation) of individuals was exchanged between the subpopulations, whereas in [59] this number depended on the subpopulation's average fitness and the individuals were selected at random. The DGA, when run on a 2-ring, 64-processor KSR1 parallel computer, showed superlinear speedup, and it took from about 4 s (32 processors) to over 100 s (1 processor) to find a solution. Unfortunately, the DGA performed differently for different parameters, and the results of the tests did not allow for unequivocal conclusions about the superiority of the DGA's over the CGA's performance. The DGA excelled the CGA on the functions *R*3 and *R*4, when migration interval $i = 50$ and migration rate $r = 0{,}5$, whereas it only achieved comparable to the CGA results on the functions *R*1 and *R*2 with $i = 5$ and 10. In the additional tests, the random migration topology implemented in the DGA did not show any superiority over hypercube or stepping stone topology, since the latter two models become equivalent to a random, island migration topology as the number of their dimensions increases. Finally, the experiments showed some inconsistence with Wright's shifting balance theory, where in contrast to the theory, the building blocks delivered by the migrants were often less fit, than those obtained as the result of the crossbreeding with the local individuals.

In [60], the potential advantage and suitability of the island model genetic algorithm in solving linearly separable (fully deceptive, Rastrigin's) and nonseparable (Powell's Singular, Rana's) parameter optimization functions were examined. There was also studied capability of the island model to maintain the diversity of the total population by using the infinite population models of simple genetic algorithms. The experiments were carried out on large (5000 individuals), and small (500 individuals) populations. The purpose of experiments carried out on the large population was to estimate the convergence behavior of a genetic algorithm using an infinite population with and without migration. The sizes of subpopulations comprising islands were set at 50, 100, 500, 1000, 5000 and the number of islands at 100, 50, 10, 5, 1 respectively. The islands were located on a counter-clockwise directed ring and the migration of individuals was performed after each 5 generations had been created on each island or after 250, 500, 2500, 5000 of fitness function evaluations per island respectively. On the first migration, a predetermined number of best individuals were copied and sent from island k to island $k - 1$, on the second migration, from k to $k - 2$ and so on, until each island had sent one set of individuals to every other island, then the process was repeated. The individuals received from other island took place of the equal number of the worst ones. In the experiments the numbers of emigrants were set at 2, 2, 5, 5 respectively. When the sizes of the total population and subpopulation were equal, the migration was not performed. The genetic algorithms were run for a maximum of 200000 evaluations on the separable functions and 400000 evaluations on the nonseparable functions. The experiments performed on large population of 5000 individuals showed that the use of migration definitely improved the performance of the Island Model genetic algorithms. In some cases, the convergence behavior of the Island Model genetic algorithms was approaching or exceeding that of a single population. In the experiments with a smaller total population size of 500 individuals, the size of population on each island was set at 50, 100, 500, the number of islands at 10, 5, 1, the migration interval at 250, 500 evaluations, and the number of emigrants 2 and 2 respectively. When the sizes of the total population and subpopulation were equal, the migration was not performed. The maximum number of fitness function evaluations carried out by the algorithms was set to 200000 evaluations or 400 generations for the separable problems and 400000 or 800 generations for the nonseparable problems. Based on the results of their experiment, the authors conclude that when migration is introduced, the performance of all Island Model genetic algorithms improved so that they performed as well or better than the single population. However, the results also suggested that Island Model genetic algorithms could still exhibit very complex and unexpected behavior, and that authors' expectations of improved performance for the Island Model on linearly separable problems was only partly correct. They also suggest that Island Model genetic algorithms consisting of 10 subpopulations of size 500 would be a reasonable design.

In [51], parallel genetic algorithm were tested on separable, non-separable, multimodal, deceptive, and epistatic problems functions to examine the influence of synchronism in the migration phase on the algorithm's performance as well as the migration frequency, search time and the speedup. The experiments were carried out on MIMD parallel computer, where identical steady-state or cellular GAs were run on 1–8 identical processors. The migration of individuals was organized on a unidirectional ring and was carried out synchronously or asynchronously. Synchronous islands waited for every incoming individual, which inevitably prolonged the PGAs' response time, while asynchronous ones did not. The migration intervals were counted in multiples of the global population size: 0μ, 1μ, 2μ, 4μ, 8μ, 16μ, 32μ, and 128μ. The migrants were chosen on the islands at random, however they were allowed to replace the worst individuals on recipient islands only if they were better than the latter ones. There was no limit imposed on the maximum number of fitness function evaluations, so, depending on the problem being solved, it took from about 50000 to 12000000 evaluations and about 3 s–850 s for PGAs to yield a solution. The results of the experiments showed that PGAs in most cases excelled the sequential GAs. In all the experiments asynchronous PGAs outperformed their synchronous equivalents in time. This is consistent with the results obtained for other PGAs and problems, and confirms the advantage of the asynchronous over synchronous interaction [61]. The tested PGAs performed with linear and sometimes with super-linear speedup when run on a cluster of workstations.

In [53], it was made a research on the effects of migration policy, which determines the way the migrants are selected and replaced, on the selection pressure and the speed of convergence in parallel multi-population evolutionary algorithms. There were considered four possible variants of selection and replacement of migrants, which were composed of random selection, selection of the best, random replacement, and replacement of the worst. It was assumed that the migrants were copied and the migration was performed after each generation, since if performed less frequently, it would have less significant effect on the convergence of the algorithm. The selection pressure was examined by determining the takeover time (which indicates how quickly a good solution, once found, occupies the entire population) and then calculating the increase in the selection intensity. The selection intensity allows for quite precise forecasting of convergence time and can be calculated analytically from the equation:

$$s^t = I \cdot \sigma_t, \tag{9.30}$$

where I – is the selection intensity, and σ_t – is the standard deviation of the population at time t, and s^t – is the selection differential, which is defined by the following equation:

$$s^t = \overline{f}_s^t - \overline{f}^t, \tag{9.31}$$

which is the difference between the mean fitness of the selected individuals s and the mean fitness of the population. The obtained experimental results show that the selection policy for choosing migrants for copying or replacement according to their fitness increases the selection pressure and can contribute to significantly faster convergence of the algorithm. It was also established that the selection pressure increases monotonically with higher migration rate, where the migration rate is assumed to be the number of individuals selected for the migration. The faster convergence can accelerate the computations and may explain some cases of superlinear speedup in parallel EAs.

In [54], an attempt was made to answer the question whether multiple independent runs of genetic algorithms (GAs) with small populations are superior to a single run with a large population, with the view of the quality of solution and the time needed to find it. The question was examined analytically under and without constraint on the number of fitness function evaluations for cases of additively-separable functions of varying difficulty. The obtained analytical and experimental results determined the situations in which the considered approaches performed best. It was established that for the functions considered, multiple runs are preferable only in conditions of limited practical value (short problems with low-order building blocks (BBs)). On the contrary, a single run with the largest population possible performs better when difficult problems (long problems with high-order BBs) were tested. Finally, the authors suggested that under the fixed cost constraint the GAs, in most cases, were able to find better solutions and in shorter time, when they were run on the largest possible population only once.

In [62], the influence of migration topology on the convergence of PGAs was investigated. Six topologies, shown in Fig. 9.1, of different interconnection densities were considered.

All graphs representing the topologies were built on 9 nodes and the migration was bidirectional in topologies A and B, and unidirectional in C–F. Topologies A and B were grids with different number of links between adjacent nodes, topologies C, E, and F were 2 (C) and 3-level hierarchical graphs converging to a single first level node with different interconnection linkage between lower levels nodes. Topology D represented a unidirectional ring. The islands in the presented topologies were either homogeneous, i.e. implementing identical GAs, or

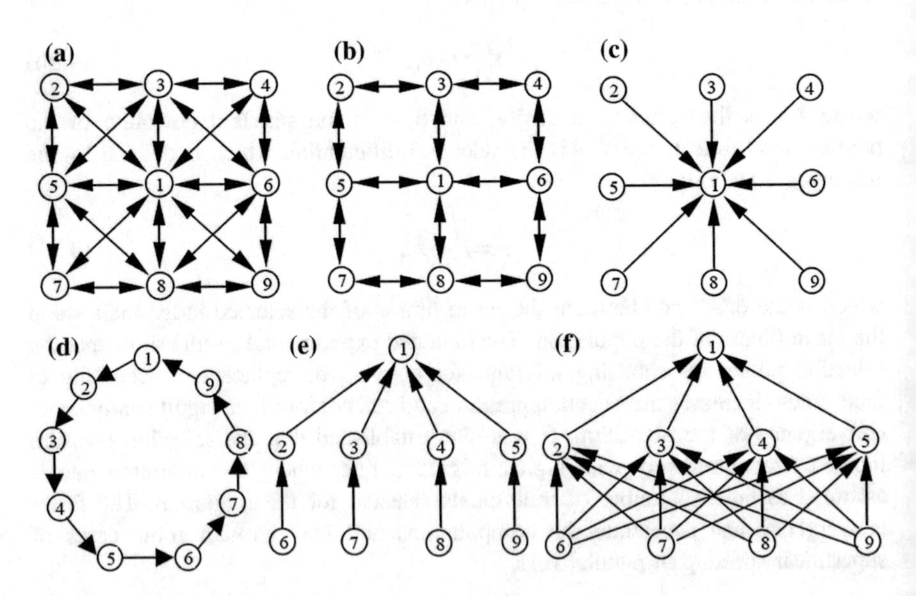

Fig. 9.1 The 6 migration topologies considered in [62]

heterogeneous, when different GAs were implemented. The migration policy assumed that a single copy of the best local individual was periodically sent from each island to the recipient island according to the chosen migration topology. The size of the population on each island was set at 20 individuals. Unfortunately, the experiment revealed that neither topology performed equally while solving the considered test problems. The topologies with dense linkage and/or frequent migration were more fit for optimization of smooth functions. In contrast, the topologies with scarce unidirectional links and rare migration were better in optimizing non-smooth (deceptive) functions. In this case, the higher degree of the diversity among the subpopulations was preserved. However, the author distinguished the hierarchical (two and three-level) topologies C, E and F as the best ones.

In [63], in contrast to spatially separated islands, an island model with islands separated in time was proposed. The basic idea underlying this concept is the reintroduction of genetic information from antecedent generations into the new one. The developed gene reuse scheme allowed to control the genetic diversity of the population with a positive effect on the convergence and performance of the algorithm. The proposed EA was composed of two island subpopulations. The first island contained the current population evolved by the EA and the second one served as a gene memory for storing the gene values (not entire chromosomes) from the antecedent populations. The gene values from the memory did not interbreed and only took part in the evolutionary process when the premature convergence had been detected. Such reintroduction of the gene values of the ancestors into the new generation backtracked and redirected the search into the new area, which allowed to maintain the diversity and the convergence on the appropriate level and, in consequence, to improve the performance of the EA. The experiments performed on two different problems, known to be prone to premature convergence, showed a significant improvement of the performance of the EA composed of islands separated in time over the basic EA.

In [64], an experimental study of the influence of migration size and migration interval on the performance and genetic diversity of the island-based EA was carried out. In the experiments, island models were composed of 2 and 5 identical islands. A standard generational EA (non-overlapping) with the exception of the best individual carried over (elitism) was assigned to each island. The total number of individuals on the archipelago was set to 100, which corresponded to the subpopulations of 50 and 20 individuals on each island respectively. The maximum number of generations allowed to be created on each island was set to 500 which corresponded to 25000 and 10000 of fitness function evaluations on each island respectively. A random-random migration policy was used, according to which migrants, selected at random from the source population, replaced randomly chosen individuals in the target population. The migration was carried out synchronously in accordance with the migration interval. In the experiments, the values of the migration interval were set to 1, 2, 3, 4, 5, 10, 15, 20, 25, 30. The migration size was varied using the values of 1, 2, 4, 6, 8, 10, 12 14, 16, 18, 19, 20. A random (dynamic) topology was used in order to provide equal relations between the islands [65].

The migration of individuals was carried out between the pairs of islands chosen at random. The performance of the island-based EA was measured with the best-so-far value from all islands. The tests were carried out using 8 functions, where four of them were specially designed to study properties of island models, and the other four were standard multimodal functions of Rosenbrock, Schwefel, Rastrigin and Griewangk. The results of the experiments revealed that the migration interval was playing much bigger role than the migration size. When the migration interval was short, the performance of the island EA had dropped noticeably. The size of migration didn't influence the behavior as much, with the exception for the sizes equal to the size of subpopulations, i.e. in the case when the migrants completely replaced the target subpopulations. For different setups the convergence took different time and in particular it took longer for rare migrations. However, in the case of rare migrations the algorithm usually converged faster with medium migration sizes, than with large or small ones. In addition, the island model very often achieved better results than the panmictic model (a single large population) and the separated islands case in which no migration was performed. There were also conducted additional experiments on the convergence of the island-based EA. The diversity was computed as standard deviation on genomes [66]. In the first experiment with fixed migration size set to 10 and variable migration interval it was observed that frequent migrations resulted in a rapid decrease of diversity. Such behavior was explained by the domination of individuals from one island over individuals from another island. When the migrations were rare, the global diversity remained high, despite the low local diversity on the individual islands. This means that different islands converged to different optima. Exchange of individuals between the islands resulted in better average fitness. In the next experiment, the convergence of the algorithm was tested for fixed migration interval set to 20 and variable migration size. It was observed that a migration of even a single individual often quickly affected the whole target population, which resulted in a diversity drop and fitness increase comparable to the case in which a large number of individuals migrate. Moreover, for small migration sizes, it was sometimes observed an increase in diversity on islands right after the migration. This suggests that a single individual was able to initiate changes in a stagnated population. The migrations of large sizes tended to have an immediate effect on the diversity on the island simply because of replacing a larger number of individuals. However, the diversity of such populations didn't change much in the following generations. In further experiments, the convergence of the algorithm with different migration intervals and sizes for the same total number of migrants was compared. For purposes of the experiment, it was assumed that migration interval was equal to migration size. When the values of these two parameters were set to 1, 2, 4, 10, 15, the algorithm converged to relatively similar fitness and showed similar diversity of population. Only, when the migration interval and the size were set to 20, the fitness and diversity were noticeably worse, compared to those obtained for previous setups. The authors finalize their discussion on the results with conclusion that the performance of the island model did not depend on a single particular parameter, but depended on the combination of them. When the combinations were appropriate, the island model outperformed both panmictic and separate setup. The best performance was obtained with moderate migration intervals and small migration sizes. Therefore, the island model algorithm would perform best with the migration interval of about 5–10 generations and the migration size less than 10% of the population. Such setup confirms earlier results reported in [67].

In [68], the discussion on the performance of the island model was continued. The dependence of performance of the island model on the amount of cooperation between the islands, the level of heterogeneity and the difficulty of the problems being solved was investigated. In the island model, composed of identical (homogeneous) islands, the spatial separation of the islands creates natural borders for the exchange of genetic information in the global population, which skillfully used may contribute to the improvement of the algorithm's performance. The diversification of the islands, i.e. making them heterogeneous, may further increase the differences between the islands and this way further enhance the desired properties of the model. Thus, it was made an attempt to find out how the heterogeneity of the islands is related to other parameters of the island model, and in which situations the heterogeneous islands are advantageous. The islands can be diversified by differentiating: EAs assigned to the islands, solution representations, fitness functions, standard parameters of EAs, or by varying the selection pressure. The introducing of heterogeneity into the island model may result in changes of certain properties of the model, e.g. may increase the maximal allowed and the minimal required level of cooperation. Thus, it was important to find out how cooperation level, the heterogeneity level and the problem difficulty influence the performance of the island model and how to choose the first two, given a specific problem to solve. A special function was designed for studying the heterogeneous island models. The function was difficult for only binary or Gray encodings, but was much easier when switching between representations was allowed [69]. For the purpose of the experiment, an island model composed of two islands was designed. One island used a standard binary encoding while the other one used a standard reflective Gray code. The model transformed individuals from one representation to another during migrations. A local optimum in one representation need not be a local optimum in the other representation. Therefore, by switching representations it was possible, in some cases, to "escape" from suboptimal solutions and to solve problems that were difficult for single representation EAs. In the experiment, the migration interval was set to 50 generations, which does not seem to confirm the supposition of the possible increase of cooperation in the heterogeneous island model. Finally, additional experiments on the robustness of the model were conducted. The island model was tested for different levels of migration using the Rosenbrock, Schwefel, Rastrigin and Griewangk functions. On the tested problems, the robustness of the island-model was comparable or better than that of the panmictic EA.

In [70], a population adapter which implements an adaptive method for determining good combinations of island population sizes and the number of islands was proposed. The adapter automatically searches for good settings of island population sizes and the number of islands according to observed performance during the run of the GA. The basic idea behind the adapter is to run three distributed GAs: DGA_0,

DGA_1, DGA_2 with the set of parameters $\{K, x\}$, $\{ K, x \cdot 2\}$, $\{K \cdot 2, x\}$ respectively, where K is the number of islands, and x is the size of the population on an island. The DGAs are allowed to run until one of the DGAs overtakes its competitors, or no further improvements are found. The overtaking is detected by comparing average fitness of DGA_0 with that of DGAs with larger total population sizes. When DGA_1 or DGA_2 overtakes DGA_0, the values of parameters K and x of DGA_0 are changed to that of the winner, and the DGA_1 or DGA_2 are restarted. A collaborative restart with seeding, which improves the efficiency of the population adapter, is implemented. In the restart with seeding, each island in the DGAs incorporates the best individuals from the other DGAs, this way setting the search towards promising regions of the search area. The population adapter was tested on the Holland's Royal Road problem and a hard real-world VLSI floor-planning problem for the cases when the restart was carried out with and without seeding. The reference value for the comparison of the test results was obtained in the experiments with manual parameter settings. The experiments with the manual settings suggested that a total population size of 2560 provides good performance of the DGA. The population adapter with restart with seeding found better parameter sets using fewer number of evaluations and with a significantly lower standard deviation than the case without seeding, which suggests that seeding leads to more reliable results. Although, the population adapter requires about three times the effort needed to solve the problem when the optimal population sizing is known beforehand, it eliminates the process of manual tuning of the considered parameters of the DGAs.

As it could be seen from the above brief review, the studying of the island model properties is the object of interest of many researches and still remains an open issue. In Chap. 11, we offer further research on the properties of the island model and propose new ideas which may contribute to its effective use.

9.8 Research on Preventing Premature Convergence in Evolutionary and Genetic Algorithms

Numerous approaches, methods, techniques and mechanisms proposed to cope with the issue of premature convergence in evolutionary and genetic algorithms can be found in the literature. A comparative review of approaches to prevent premature convergence in GA is given in [50] and an overview of methods maintaining diversity in genetic algorithms in [71]. In addition to the empirical work, there has also been conducted theoretical research on population diversification.

In [72], the global exploration capabilities of mutation-based algorithms for sub-linear population sizes were investigated theoretically and empirically. Using a simple bimodal test function and rigorous runtime analyses, the authors compared well-known genotype and fitness diversity preserving mechanisms, deterministic crowding and fitness sharing with a plain $(\mu + 1)$ EA without diversification. It was shown that diversification is necessary for global exploration, but not all mechanisms succeed to the same extent. According to the authors, population is nearly useless for the algorithms with genotype and fitness diversity, as it was experienced the same performance as for simple hill climbers like local search or the (1 + 1) EA. On the other hand, algorithms with fitness sharing and deterministic crowding performed better than with genotype and fitness diversity due to their higher success probabilities. Their experimental results indicate a similar behavior of the diversity preserving mechanisms also for larger populations.

In [73], theoretical analysis and empirical study on the $(\mu + 1)$ EA operating on the Balance Function was conducted. It follows from the experiments that linear population sizes are sufficient to make both the $(\mu + 1)$ EA without diversity and equipped with genotype diversity efficient for Balance function. Also there appeared to be a sharp threshold at population size cn for some $c > 0$ at which the expected performance of these two algorithms turns from exponential to polynomial. On the other hand, the $(\mu + 1)$ EA with fitness diversity appears to be effective for population sizes considerably smaller than the sublinear ones required for their proof to work. The authors stated also, that for larger population sizes the crowding mechanism becomes detrimental.

It follows from the last two cited works that the effectiveness of EA, depends not only on the diversification mechanisms used, but also on the size of the population evolved. It seems that the authors of these two works differently conclude on the effectiveness of genotype, fitness, and deterministic crowding mechanisms. For example, according to [72] the algorithms without diversity or with genotype or fitness diversity preserving mechanisms were considered as weak and performed worse than with deterministic crowding, which is not entirely consistent with the results obtained in [73]. Such inconsistency testifies to the fact that the efficiency of diversification mechanisms still remains an open issue and requires further theoretical and empirical research.

In [74], a Social Disasters Technique (SDT) was proposed, where packing catastrophic operator was used for fitness diversity preserving by replacing individuals of the same fitness, except for one, with random individuals in the situations when the level of population diversity dropped to some preset limit.

In [75] it was proposed a Random Offspring Generation (ROG) technique, aimed at genotype diversity preserving, by which the crossover is carried out only on parents with different genotypes, otherwise one or two offspring were generated at random. Although this technique prevents the formation of clone-offspring from clones-parents, it does not prevent the transition of clones from generation to generation, since each individual with an old population has the right to go to the next one with probability proportional to its quality.

In [72] it was proposed $(\mu + 1)$ EA with genotype diversity, which was based on (2 + 1) GA, described in [76]. This algorithm does not allow an offspring to enter the population, if there already exists its genetic duplicate.

References

1. Józefowska, J., Węglarz, J.: On a methodology for discrete-continuous scheduling. Eur. J. Oper. Res. **107**(2), 338–353 (1998)
2. Gorczyca, M., Janiak, A.: New approach to resource allocation in the problems of scheduling of tasks described with concave dynamic models. In: Kaszyński, R. (ed.) Proceedings of the 13-th IEEE/IFAC International Conference on Methods and Models in Automation and Robotics, Szczecin, pp. 1189–1192 (2007)
3. Gorczyca, M., Janiak, A.: Methods for the optimal resource allocation in the problems of scheduling of tasks described with concave dynamic model. In: 14th IFAC Conference on Methods and Models in Automation and Robotics, IFAC Proceedings, vol. 42(13), pp. 250–255 (2009)
4. Gorczyca, M.: Resource allocation and task scheduling algorithms for the selected problems with dynamic task models and parallel processors. Ph.D. thesis (in Polish), Wrocław University of Technology (2008)
5. Gorczyca, M., Janiak, A.: Resource level minimization in the discrete–continuous scheduling. Eur. J. Oper. Res. **203**, 32–41 (2010)
6. Schittowski, K.: NLQPL: a FORTRAN-subroutine solving constrained nonlinear programming problems. Ann. Oper. Res. **5**, 485–500 (1985)
7. Gorczyca, M., Janiak, A.: Dominance properties in the discrete-continuous scheduling problems, International Conference on System Science, Wrocław, pp. 96–106 (2007)
8. Gorczyca, M., Janiak, A., Janiak, W.: The discrete part of the discrete-continuous scheduling problems—new properties. In: 14th IFAC Conference on Methods and Models in Automation and Robotics, IFAC Proceedigns, vol. 42(13), pp 244–249 (2009)
9. Józefowska, J., Mika, M., Różycki, R., Waligóra, G., Węglarz, J.: Solving the discrete-continuous project scheduling problem via its discretization. Math. Methods Oper. Res. **52**(3), 489–499 (2000)
10. Różycki, R.: Zastosowanie algorytmu genetycznego do rozwiązywania dyskretno-ciągłych problemów szeregowania. Ph.D. dissertation, Poznań University of Technology, Poland (2000)
11. Józefowska, J., Mika, M., Różycki, R., Waligóra, G., Węglarz, J.: A heuristic approach to allocating the continuous resource in discrete–continuous scheduling problems to minimize the makespan. J. Sched. **5**(6), 487–499 (2002)
12. Józefowska, J., Waligóra, G.: Heuristic procedures for allocating the continuous resource in discrete-continuous scheduling problems. Found. Comput. Decis. Sci. **29**(4), 315–328 (2004)
13. Waligóra, G.: Tabu search for discrete-continuous scheduling problems with heuristic continuous resource allocation. Eur. J. Oper. Res. **193**(3), 849–856 (2009)
14. Waligóra, G.: Heuristic approaches to discrete-continuous project scheduling problems to minimize the makespan. Comput. Optim. Appl. **48**(2), 399–421 (2011)
15. Jędrzejowicz, P., Skakovski, A.: An Island-Based Evolution Algorithm for Discrete-Continuous Scheduling with Continuous Resource Discretisation, Proceedings of the 2nd IEEE International Conference on Computational Cybernetics ICCC 2004, Aug 30–Sept 1, 2004, Vienna University of Technology, Austria (2004)
16. Jędrzejowicz, P., Skakovski, A.: A population learning algorithm for discrete-continuous scheduling with continuous resource discretisation. In: Chen, Y., Abraham, A., Jinan, B. (eds.) Proceedings of 6th International Conference on Intelligent Systems Design and Applications (ISDA 2006), vol. 2, spec. sess.: Nature Imitation Methods Theory and practice (NIM'06), Peoples Republic of China, pp. 1153–1158 (2006)
17. Jędrzejowicz, P., Skakovski, A.: A cross-entropy based population learning algorithm for discrete-continuous scheduling with continuous resource discretisation. Neurocomputing 73 (4–6), Special Issue: SI, 655–660 (2010)

18. Jędrzejowicz, P., Skakovski, A.: Structure versus efficiency of the cross-entropy based population learning algorithm for discrete-continuous scheduling with continuous resource discretisation. In: Czarnowski, I., Jędrzejowicz, P., Kacprzyk, J. (eds.) Studies in Computational Intelligence. Agent-Based Optimization, vol. 456, pp. 77–102 (2013)
19. Jędrzejowicz, P., Skakovski, A.: Population learning with differential evolution for the discrete-continuous scheduling with continuous resource discretisation. In: IEEE International Conference on Cybernetics (CYBCONF) Lausanne, Switzerland, 13–15 June, pp. 92–97 (2013)
20. Jędrzejowicz, P., Skakovski, A.: Island-based differential evolution algorithm for the discrete-continuous scheduling with continuous resource discretisation. Procedia Comput. Sci. **35**, 111–117 (2014)
21. Jędrzejowicz, P., Skakovski, A.: Improving performance of the differential evolution algorithm using cyclic decloning and changeable population size. In: Nguyen, N.T., Czarnowski, I., Hwang, D. (eds.), Journal of Universal Computer Science (J.UCS), Special Issue—Computational Intelligence Tools for Processing Collective Data (CITPCD 15), vol. 22 (6), pp. 874–893 (2016)
22. Jędrzejowicz, P., Skakovski, A.: Properties of the Island-Based and single population differential evolution algorithms applied to discrete-continuous scheduling. In: Czarnowski, I. et al. (eds.) Intelligent Decision Technologies 2016, Proceedings of the 8th KES International Conference on Intelligent Decision Technologies (KES-IDT 2016)—Part I, Smart Innovation, Systems and Technologies, vol. 56, pp. 349–359 (2016)
23. Słowiński, R.: Algorytmy sterowania rozdziałem zasobów różnych kategorii w kompleksie operacji. Wydawnictwo Politechniki Poznańskiej, seria Rozprawy Nr 114, Poznań (1980)
24. Drexl, A., Gruenewald, J.: Nonpreemptive multi-mode resource-constrained project scheduling. IIE Trans. **25**(5), 74–81 (1993)
25. Hartmann, S.: Project scheduling with multiple modes: a genetic algorithm (in English). Manuskripte aus den Instituten für Betriebswirtschaftslehre Nr. 435, the University of Kiel, Germany (1997)
26. Hartmann, S., Briskorn, D.: A survey of variants and extensions of the resource-constrained project scheduling problem. Eur. J. Oper. Res. **207**(1), 1–14 (2010)
27. Bartusch, M., Rolf, H.M., Radermacher, F.J.: Scheduling project networks with resource constraints and time windows. Ann. Oper. Res. **16**, 201–240 (1988)
28. Józefowska, J., Mika, M., Węglarz, J.: A simulated annealing algorithm for some class of discrete-continuous scheduling problems. Computational Methods in Science and Technology, vol. 2(1), pp. 73–85. Scientific Publishers OWN, Poznan (1996)
29. Blazewicz, J., Kubiak, W., Szwarcfiter, J.: Scheduling independent fixed-type tasks. In: Słowiński, R., Węglarz, J. (eds.) Advances in Project Scheduling. Elsevier, Amsterdam (1989)
30. Józefowska, J., Mika, M., Różycki, R., Waligóra, G., Węglarz, J.: Discrete-continuous scheduling to minimize the makespan for power processing rates of jobs. Discret. Appl. Math. **94**, 263–285 (1999)
31. Lawrence, C., Zhou, J.L., Tits, A.L.: Users guide for CFSQP Version 2.3. Available by e-mail: andre@ eng.umd.edu (1995)
32. Lawrence, C., Zhou, J.L., Tits, A.L.: Users guide for CFSQP Version 2.5. Available by email: andre@eng.umd.edu (1997)
33. Józefowska, J., Różycki, R., Waligóra, G., Węglarz, J.: Local search metaheuristics for discrete-continuous scheduling problems. Eur. J Oper. Res. **107**(2), 354–370 (1998)
34. Józefowska, J., Mika, M., Różycki, R., Waligóra, G., Węglarz, J.: Discrete-Continuous scheduling to minimize the mean flow time—computational experiments. Comput. Methods Sci Technol. **3**(1), 25–37 (1997)
35. Józefowska, J., Mika, M., Różycki, R., Waligóra, G., Węglarz, J.: Discrete-continuous scheduling to minimize maximum lateness. In: Proceedings of the Fourth International Symposium on Methods and Models in Automation and Robotics MMAR'97, Międzyzdroje 26–29 Aug 1997, pp. 947–952 (1997)

36. Józefowska, J., Mika, M., Różycki, R., Waligóra, G., Węglarz, J.: Solving discrete-continuous scheduling problems by Tabu Search. In: 4th Metaheuristics International Conference MIC'2001, Porto, Portugal, 16–20 July 2001, pp. 667–671 (2001)
37. Józefowska, J., Waligóra, G., Węglarz, J.: Tabu list management methods for a discrete–continuous scheduling problem. Eur. J. Oper. Res. **137**, 288–302 (2002)
38. Skorin-Kapov, J.: Tabu search applied to the quadratic assignment problem. ORSA J. Comput. **2**, 33–45 (1990)
39. Glover, F.: Tabu search- Part 1. ORSA J. Comput. **1**, 190–206 (1989)
40. Glover, F., Laguna, M.: Tabu Search. Kluwer Academic Publishers, Norwell (1997)
41. Józefowska, J., Waligóra, G., Węglarz, J.: A Performance Analysis of Tabu Search for Discrete-Continuous Scheduling Problems. Metaheuristics: Computer Decision-Making, pp. 385–404. Kluwer Academic Publishers B. V. (2003)
42. Janiak, A.: Minimization of the blooming mill standstills—mathematical model. Suboptimal algorithms. Zesz. Nauk. AGH, s. Mechanika **8**(2), 37–49 (1989)
43. Kurts, D.S., Swartz, C.W.: Theories of Integration. World Scientific (2004)
44. Różycki, R., Węglarz, J.: On job models in power management problems. Bull. Pol. Acad. Sci. Tech. Sci. **57**(2), 147–151 (2009)
45. Li, M., Yao, A.C., Yao, F.F.: Discrete and continuous min-energy schedules for variable voltage processors. In: Proceedings of the National Academy of Sciences of the USA, vol. 103 (11), pp. 3983–3987 (2006)
46. Yao, F., Demers, A., Shenker, S.: A scheduling model for reduced CPU energy. In: Proceedings of the 36th IEEE Conference on the Foundations of Computer Science (FOCS) (IEEE, New York), pp. 374–382 (1995)
47. Kwon, W., Kim, T.: Optimal voltage allocation techniques for dynamically variable voltage processors. ACM Trans. Embed. Comput. Syst. **4**(1), 211–230 (2005)
48. Li, M., Yao, F.F.: An efficient algorithm for computing optimal discrete voltage schedules. SIAM J. Comput. **35**(3), 658–671 (2006)
49. Brinkmann, A., Kling, P., Meyer auf der Heide, F., Nagel, L., Riechers, S., Süß, T.: Scheduling shared continuous resources on many-cores. In: Proceedings of the 26th ACM Symposium on Parallelism in Algorithms and Architectures SPAA '14, Prague, Czech Republic, June 23–25, pp. 128–137 (2014)
50. Pandey, H.M., Chaudharyb, A., Mehrotra, D.: A comparative review of approaches to prevent premature convergence in GA. Appl. Soft Comput. **24**, 1047–1077 (2014)
51. Alba, E., Troya, J.: Analysis of synchronous and asynchronous parallel distributed genetic algorithms with structured and panmictic Islands. In: Rolim, J., et al. (eds.) Proceedings of the 10th Symposium on Parallel and Distributed Processing. San Juan, Puerto Rico, USA, 12–16 Aprl, pp. 248–256 (1999)
52. Belding, T.C.: The distributed genetic algorithm revisited. In: Eshelman, L.J. (ed.) Proceedings of the Sixth International Conference on Genetic Algorithms, pp. 114–121. Morgan Kaufmann, San Francisco CA (1995)
53. Cantu-Paz, E.: Migration policies, selection pressure, and parallel evolutionary algorithms. J. Heuristics **7**(4), 31–334 (2001)
54. Cantu-Paz, E., Goldberg, D.E.: Are multiple runs of genetic algorithms better than one? In: Proceedings of the Genetic and Evolutionary Computation Conference (2003)
55. Muhlenbein, H.: Evolution in time and space: the parallel genetic algorithm. In: Rawlins, G. (ed.) FOGA-1,. pp. 316–337. Morgan Kaufman (1991)
56. Whitley, D., Starkweather, T.: GENITOR II: a distributed genetic algorithm. J. Exp. Theor. Artif. Intell. **2**(3), 33–47 (1990)
57. Wright, S.: Evolution in mendelian populations. Genetics **16**, 97–159 (1931)
58. Wright, S.: Isolation by distance. Genetics **28**, 114–138 (1943)
59. Tanese, R.: Parallel genetic algorithms for a hypercube. In: Grefenstette, J.J. (ed.) Hillsdale, pp. 177–183. Lawrence Erlbaum, NJ (1987)
60. Whitley, D., Rana, S., Heckendorn, R.B.: The island model genetic algorithm: on separability, population size and convergence. J. Comput. Inf. Technol. **7**(1), 33–47 (1999)

61. Hart, W.E., Baden, S., Belew, R.K., Kohn, S.: Analysis of the numerical effects of parallelism on a parallel genetic algorithm. In: IEEE (ed.): CD-ROM IPPS97 (1997)
62. Sekaj, I.: Robust parallel genetic algorithms with re-initialisation. In: Proceedings of Parallel Problem Solving from Nature—PPSN VIII, 8th International Conference, Birmingham, UK, Sept 18–22, LNCS, vol. 3242, pp. 411–419. Springer (2004)
63. Prime, B., Hendtlass, T.: Evolutionary Computation Using Island Populations in Time. Innovations in Applied Artificial Intelligence, LNCS 3029, pp. 573–582 (2004)
64. Skolicki, Z., Kenneth, D.J.: The influence of migration sizes and intervals on island models. In: Proceedings of GECCO'05, June 25–29, Washington, DC, USA, pp. 1295–1302 (2005)
65. de Vega, F.F., Tomassini, M., Punch III, W.F., Sanchez-Prez, J.M.: Experimental study of multipopulation parallel genetic programming. In: Proceedings of the European Conference on Genetic Programming, Lecture Notes in Computer Science, vol. 1802, pp. 283–293. Springer (2000)
66. Morrison, R. W.: Designing evolutionary algorithms for dynamic environments. Natural Computing Series. Springer (2004)
67. Tomassini, M.: Spatially structured EAs. In: GECCO'04 Tutorials, June 2004
68. Skolicki, Z.: An analysis of Island models in evolutionary computation. In: Proceedings of GECCO'05, June 25–29, Washington, DC, USA, pp. 386–389 (2005)
69. Skolicki, Z., Kenneth, D.J.: Improving evolutionary algorithms with multi-representation island models. In: Parallel Problem Solving from Nature—PPSN VIII, LNCS 3242, pp. 420–429 (2004)
70. Berntsson, J., Tang, M.: Adaptive sizing of populations and number of Islands in distributed genetic algorithms. In: Proceedings of 2005 Genetic and Evolutionary Computation Conference GECCO'05, ACM, pp. 1575–1576 (2005)
71. Gupta, D., Ghafir, S.: An overview of methods maintaining diversity in genetic algorithms. Int. J. Emer. Technol. Adv. Eng. 2, 5 (2012). https://www.ijetae.com
72. Friedrich, T., Oliveto, P.S., Sudholt, D., Witt, C.: Analysis of diversity-preserving mechanisms for global exploration. Evol. Comput. **17**(4), 455–476 (2009)
73. Oliveto, P.S., Zarges, C.: Analysis of diversity mechanisms for optimisation in dynamic environments with low frequencies of change. Theor. Comput. Sci. 561(A), pp. 37–56 (2015)
74. Kureichick, V.M., Melikhov, A.N., Miaghick, V.V., Savelev, O.V., Topchy, A.P.: Some new features in the genetic solution of the traveling salesman problem. In: Proceedings of ACEDC'96, Plymouth (1996)
75. Rocha, M., Neves, J.: Preventing Premature Convergence to Local Optima in Genetic Algorithms via Random Offspring Generation; LNAI (Lecture Notes in Artificial Intelligence), vol. 1611, pp. 127–136 (1999)
76. Storch, T., Wegener, I.: Real royal road functions for constant population size. Theoret. Comput. Sci. **320**(1), 123–134 (2004)

Chapter 10
Proposed Metaheuristics for Solving Problem Θ_Z (DCSPwCRD)

10.1 IBEA—Island-Based Evolutionary Algorithm

An island-based evolution algorithm (IBEA) belongs to the class of distributed algorithms. To improve efficiency of genetic algorithms (GA) several distributed GA's were proposed in [1–3]. The proposed algorithms used an island-based approach where a set of independent populations of individuals evolves on "islands" cooperating with each other. The island-based approach brings two benefits: a model that maps easily onto the parallel hardware and extended search area (due to multiplicity of islands) preventing from sticking in local optima. Promising results of the island-based approach achieved in [4, 5] motivated the author to design the IBEA for discrete-continuous scheduling.

An island-based evolutionary algorithm (IBEA), proposed originally in [6], operates on two levels: on the island level and population level. To evolve individuals of the population level a population-based evolutionary algorithm (PBEA) is proposed. On the island level the following assumptions are made:

- all islands are located on a directed ring,
- an island is represented by a population of individuals,
- the populations of individuals evolve on each island independently,
- each island I_k regularly sends its best solution to the successor $I_{(k \ mod \ K) + 1}$ in the ring, where $k = 1, 2, \ldots, K$, and K is the number of islands,

On the population level the following assumptions are made:

- an individual (a solution) is represented by an n-element vector $S = [c_i| \ 1 \leq i \leq n]$,

E. Ratajczak-Ropel and A. Skakovski, *Population-Based Approaches to the Resource-Constrained and Discrete-Continuous Scheduling*, Studies in Systems, Decision and Control 108, DOI 10.1007/978-3-319-62893-6_10

- all processing modes of all tasks are numbered consecutively. Thus, processing mode l_b of task J_b has the number $c_b = \sum_{i=1}^{b-1} W_i + l_b$,
- all S representing feasible solutions are potential individuals,
- an initial population P_0 is composed from the potential individuals for whom task modes and tasks order on the list are random,
- each individual can be transformed into a schedule by applying LSG, which is a specially designed list-scheduling algorithm for discrete-continuous scheduling,
- each schedule produced by the LSG can be directly evaluated in terms of its fitness,
- new population is formed by applying several evolution operators: selection and transfer of some more "fit" individuals, random generation of individuals, crossover, and mutation,
- the algorithm stops when an optimality criterion is satisfied or the preset number of generations on each island have been generated,
- when IBEA stops, the best overall solution is the final one.

The following pseudo-code shows main stages of the IBEA algorithm:

```
Procedure IBEA
Begin
 Set number of islands K, number of generations PN and
 size of the population PS for each island.
 For each island I_k, generate an initial population P_0.
 While no stopping criteria is met do
  For each island I_k do
   Evolve PN generations on island I_k using PBEA.
   Send the best solution to I_(k mod K) + 1.
   Incorporate the best solution from I_((K+k-2) mod K)+1 in-
   stead of the best one.
 EndWhile.
 Find the best solution S_best across all islands and
 save it as the final one.
EndProc_IBEA.
```

The PBEA algorithm is shown in the following pseudo-code:

```
Procedure PBEA
Begin
 Set ic:= 0; (ic - iteration counter);
 While no stopping criteria is met do
  Set ic:= ic + 1
  Calculate fitness factor for each individual in pop-
  ulation P_{ic-1} using LSG;
  Form new population P_{ic}:
  Select randomly a quarter of PS of individuals from
  P_{ic-1} (probability of selection depends on fitness of
  an individual);
  Create a quarter of PS of individuals by applying
  crossover operator to previously selected individu-
  als from P_{ic-1};
  Create a quarter of PS of individuals by applying
  mutation operators to previously selected individu-
  als from P_{ic-1};
  Generate a quarter of PS of individuals from the
  set of potential individuals (random task pro-
  cessing mode, and task order);
 EndWhile.
EndProc_PBEA.
```

The LSG algorithm used to transform S into a schedule is carried out as follows:

```
Procedure LSG
Begin
 Construct a list of tasks from the code representing
 an individual. Set loop over tasks on the list.
 Within the loop, allocate current task to a processor
 considering the amount of a continuous resource allo-
 cated to the task, and minimizing the beginning time
 of its processing. Continue with tasks until all have
 been allocated.
 Determine the fitness of individual S as Q_s = max{C_i},
 i = 1, ... , n.
EndProc_LSG.
```

Table 10.1 The comparison of the results obtained by the IBEA and G_{dskr} for problem Θ_Z

Problem size n × m × W	Relative error (RE)	Discretisation level					
		$W = 10$		$W = 20$		$W = 50$	
		IBEA (%)	Gdskr (%)	IBEA (%)	G_{dskr} (%)	IBEA (%)	Gdskr (%)
$10 \times 2 \times W$	RE_{avg}	−3,09	7,59	−2,77	9,18	−2,86	10,41
	RE_{max}	50,02	15,34	49,09	15,84	52,22	17,54
$10 \times 3 \times W$	RE_{avg}	−0,45	14,58	−0,98	15,79	−0,14	17,11
	RE_{max}	73,33	25,18	69,65	23,15	71,91	27,52
$20 \times 2 \times W$	RE_{avg}	4,08	13,25	4,74	15,42	4,91	17,65
	RE_{max}	41,09	19,40	43,44	24,02	43,28	26,87

10.1.1 Computational Experiment

The proposed island-based evolution algorithm for solving discrete-continuous scheduling problems with continuous resource discretisation was implemented and tested. Results were compared to the best known obtained by a genetic GAVR-dyskr, tabu search, and simulated annealing algorithms [7] (GAVRdyskr, denoted as G_{dskr}, was used for results comparison as the one of the same nature).

Three combinations of $n \times m$ were considered (n—the number of tasks and m—the number of machines): 10×2, 10×3, and 20×2. For each $n \times m$ combination three discretisation levels were considered: 10, 20, and 50. For each discretisation level 100 instances of a problem Θ_Z were generated, which makes 900 instances of the problem. Each instance was tested 20 times. Relative error (RE) of the solutions found by the IBEA compared to best-known solutions was used to evaluate the quality of IBEA. RE calculated as $RE = (Q_{IBEA} - Q_{best\text{-}known})/Q_{best\text{-}known}$ for each instance was used to find average (RE_{avg}) and maximum (RE_{max}) relative errors. RE_{avg} and RE_{max} of the solutions found by the IBEA and G_{dskr} are presented in Table 10.1.

As it can be seen in Table 10.1 the quality of the solutions found by the IBEA is, on average, competitive with the quality of the solutions found by G_{dskr}. For example, for case 10×2, $W = 10$ $RE_{avg} = -3{,}09\%$, which means that the schedule length of all schedules yielded by IBEA was 3,09% shorter on average than the best-known. For the same case, $RE_{max} = 50{,}02\%$ means that the longest schedule among all schedules yielded by IBEA was 50,02% longer than the best-known. Such large RE_{max} points to the necessity of better tunning of the IBEA. As it can be also seen from Table 10.1, REs of the solutions do not always decrease as continuous resource discretisation level increases. Thus the level of continuous resource discretisation for which REs of the solutions are smallest should be determined empirically.

Mean time required by the IBEA to find a solution for 10 × 2 on CPU with Pentium III 733 MHz was 60 s. Mean time required by G_{dskr} to find a solution for 10 × 2 on supercomputer Silicon Graphics Power Challenge XL with twelve RISC MIPS R8000 processors was 33 s.

Such results make IBEA quite effective algorithm for solving discrete-continuous scheduling problems with continuous resource discretisation.

10.2 PLA—Population Learning Algorithm

Population learning algorithm (PLA) first introduced in [8] takes advantage of basic ideas, principals and assumptions introduced in a Social Learning Algorithm (SLA) originally proposed in [9]. Both SLA and PLA are based on an analogy to a social phenomenon rather than to evolutionary processes. Whereas evolutionary algorithms emulate basic features of natural evolution including natural selection, hereditary variations, the survival of the fittest and production of far more offspring than are necessary to replace current generation, the population learning algorithms take advantage of features that are common to social education systems:

- a generation of individuals enters the system,
- individuals learn through organized tuition, interaction, self-study and self-improvement,
- learning process is inherently parallel with different schools, curricula, teachers, etc.,
- learning process is divided into stages,
- more advanced and more demanding stages are entered by a diminishing number of individuals from the initial population (generation),
- at higher stages more advanced education techniques are used,
- the final stage can be reached by only a fraction of the initial population.

In the PLA, the assumptions made for the individuals are the same as the assumptions made for the individuals of the IBEA on the population level, see Sect. 10.1.

Initially, a number of individuals, known as the initial population, is randomly generated. Once the initial population has been generated, individuals enter the first learning stage. It involves applying some, possibly basic and elementary, improvement schemes. These can be based, for example, on some simple local search procedures. The improved individuals are then evaluated and better ones pass to the subsequent stage. A strategy of selecting better or more promising individuals must be defined and duly applied. In the following stages the whole cycle is repeated. Individuals are subject to improvement and learning, either individually or through information exchange, and the selected ones are again promoted to the higher stage with the remaining ones dropped-out from the process. At the final stage the remaining individuals are reviewed and the best represents a solution to the problem at hand.

The PLA is seen here as a general framework for constructing hybrid solutions to difficult computational problems. Strength of the PLA stems from integrating in an "intelligent" manner the power of population-based algorithms using some random mechanism for diversity assurance, with efficiency of various local search algorithms. The later may include, for example, reactive search, tabu search, simulated annealing as well as the described earlier population based approaches.

General idea of the present implementation of the PLA proposed in [8] is shown in the following pseudo code:

```
Procedure PLA
Begin
  Generate an initial population P0 of size x0, where
  task order and modes in each individual are random.
  Select x1 most promising individuals from P0 to P1,
  where x1 << x0.
  Learn all individuals in P1 with aid of procedure
  IBEA.
  Select the best individual (BI) from P1 to the last
  learning stage.
  Learn BI with the aid of procedure TS.
  Output the best solution to the problem.
EndProc_PLA.
```

In the presented pseudo code, procedure IBEA stands for the Island-Based Evolution Algorithm described in Sect. 10.1, and TS—for Tabu Search, which is to be presented in the subsequent section.

10.2.1 Tabu Search

Tabu search is another metaheuristic used in the considered PLA (see [10]). In order to present the general idea of present implementation of the tabu search procedure, we introduce the neighborhoods N_t and N_{md} of a solution S. N_t is a set of solutions generated from S by moving task $J_i \in S$ from place i to the rest $n - 1$ places. Thus, we yield $|N_t| = n \cdot (n - 1)$ neighbors. N_{md} is a set of solutions generated from S by assigning to task $J_i \in S$ one by one in a row all of its W modes, assuming that all tasks can be executed in W modes. Thus we yield another $|N_{md}| = n \cdot (W - 1)$ neighbors. The considered Tabu Search procedure is shown in the following pseudo code:

```
Procedure TS
Begin
  Set S0 = SI, where SI = BI from P1.
  Set the best solution Sbest = S0.
  Set Tabu List TL = {∅}.
  Set Nt = {SI} and Nmd = {∅}.
  Repeat the following max_number_of_iterations times:
    Find the best legal neighbour Sbln of S0, i.e. the
    best across Nt and Nmd neighbour which is not on TL.
    Set S0 = Sbln.
    If Sbln is more fit than Sbest then Sbest = Sbln.
    Put Sbln on the Tabu list.
    If the fitness of S0 has not improved after nit num-
    ber of iterations construct a new solution by moving
    task Ji in S0 to one of the chosen randomly less fre-
    quently visited places on the task list and assign-
    ing to it one of the chosen randomly less frequently
    assigned execution modes.
  EndRepeat.
EndProc_TS.
```

The size of the Tabu List (TL) was determined empirically and set to 500 solutions.

10.2.2 Computational Experiment

The proposed population learning algorithm for solving discrete-continuous scheduling problems with continuous resource discretisation was implemented and tested. Results were compared to the best known obtained by a genetic GAVRdyskr, tabu search, simulated annealing algorithms [7], and the IBEA described in [4] (GAVRdyskr, denoted in the rest of the text as G_{dskr}, and the IBEA were used for results comparison in as the ones of the same nature). For testing purposes three combinations of $n \times m$ were considered (n—the number of tasks and m—the number of machines): 10×2, 10×3, and 20×2. For each $n \times m$ combination three discretisation levels were considered: 10, 20, and 50. For each discretisation level 100 instances of a problem Θ_Z were generated, which makes 900 instances of the problem. The instances of the problem were generated with aid of a procedure received from the author of [7]. Each instance was tested 26 times. Relative error (RE) of the solutions found by the PLA compared to best-known solutions was used to evaluate the quality of the PLA.

The value of the RE calculated for each instance according to the formulae $RE = (Q_{PLA} - Q_{best\text{-}known})/Q_{best\text{-}known}$ was used to find average (RE_{avg}) and maximum (RE_{max}) relative errors. RE_{avg} and RE_{max} of the solutions found by the PLA, IBEA, and G_{dskr} are presented in Table 10.2. As it can be seen in Table 10.2, the quality of the solutions found by the PLA has been, on average, better than the quality of the solutions found by the IBEA and G_{dskr}. For example, for case 10×2, $W = 10$, $RE_{avg} = -2{,}12\%$, which means that the schedule length of all schedules yielded by the PLA was 2,12% on average shorter than the best-known. For the same case, $RE_{max} = 56{,}34\%$ means that the longest schedule among all schedules yielded by PLA was 56,34% longer than the best-known. Such large RE_{max} points to the necessity of better tuning of the PLA. Despite large RE_{max}, the PLA improved 55% of 300 the best known solutions for sizes 10×2, 10×3 and 20×2, and reduced average values of RE_{avg} compared to the IBEA and G_{dskr}. Table 10.3 shows the average of the PLA's RE_{avg} reduction percentage in comparison to average RE_{avg} of the IBEA and G_{dskr}. In other words, Table 10.3 shows how many percent on average the solutions found by the PLA were better than the solutions found by the IBEA and G_{dskr}. As it can be also seen from Table 10.2, REs of the solutions do not always decrease as continuous resource discretisation level increases. Thus, the level of continuous resource discretisation for which REs of the solutions are smallest should be determined empirically.

Table 10.2 The comparison of REs obtained by the PLA, IBEA and G_{dskr} for problem Θ_Z

Problem size $n \times m \times W$	Relative error (RE)	Discretisation level								
		$W = 10$			$W = 20$			$W = 50$		
		PLA (%)	IBEA (%)	G_{dskr} (%)	PLA (%)	IBEA (%)	G_{dskr} (%)	PLA (%)	IBEA (%)	G_{dskr} (%)
$10 \times 2 \times W$	RE_{avg}	−2,12	−0,61	7,59	−1,44	0,19	9,18	−1,07	0,36	10,41
	RE_{max}	56,34	62,35	15,34	56,69	62,42	15,84	61,55	62,88	17,54
$10 \times 3 \times W$	RE_{avg}	1,46	2,89	14,58	1,03	3,49	15,79	1,66	5,22	17,11
	RE_{max}	77,51	83,11	25,18	76,17	84,91	23,15	78,80	82,40	27,52
$20 \times 2 \times W$	RE_{avg}	5,62	6,62	13,25	6,27	6,96	15,42	6,59	7,13	17,65
	RE_{max}	46,50	47,84	19,40	47,60	47,8	24,02	47,83	47,86	26,87

Table 10.3 The average of the PLA's RE_{avg} reduction percentage in comparison to average RE_{avg} of the IBEA and G_{dskr}

Problem size $n \times m \times W$	IBEA (%)	G_{dskr} (%)
$10 \times 2 \times W$	1,52	10,60
$10 \times 3 \times W$	2,48	14,44
$20 \times 2 \times W$	0,74	9,28

Mean time required by the PLA to find a solution for 10×2 on Pentium (R) 4 CPU 3.00 GHz compiled with aid of Borland Delphi Personal v.7.0 was 5 s, by IBEA compiled in Borland Pascal v.7.0—48 s. Mean time required by G_{dskr} to find a solution for 10×2 on supercomputer Silicon Graphics Power Challenge XL with twelve RISC MIPS R8000 processors was 33 s. Such results make PLA quite effective algorithm for solving discrete-continuous scheduling problems with continuous resource discretisation.

10.3 PLA2—Cross-Entropy-Based Population Learning Algorithm

A cross-entropy-based population learning algorithm (PLA2) proposed in [11] is another attempt to make use of the idea of social learning framework already presented in Sect. 10.2. Different to the PLA structure, more advanced procedure for the initial population creation and different setting for the TS procedure contributed to higher efficiency of the PLA2.

In the PLA2, the assumptions made for the individuals are the same as the assumptions made for the individuals of the IBEA on the population level, see Sect. 10.1. The general idea of the implementation of the PLA2 is shown in the following pseudo code:

```
Procedure PLA2
Begin
  Create an initial population P0 of the size x0 - 1 us-
  ing procedure cross-entropy (CE).
  Create an individual TSI in which all tasks Ji are to
  be executed in mode li = 1 (a mode characterized by
  minimal quantity of additional resource u_i^1 and maxi-
  mal task processing time τ_i^1, 1 ≤ i ≤ n).
  Improve individual TSI with Tabu Search (TS) proce-
  dure.
  Create population P1 = P0 + TSI.
  Improve all individuals in P1 with IBEA.
  Output the best solution to the problem.
EndProc_PLA2.
```

In the description of the procedure PLA2 above, $x_0 = K \cdot PS$, where *K*—the number of islands and *PS*—the population size on an island defined in procedure IBEA. As it follows from the description of the PLA2, population P_1 comprises the initial population for the IBEA, therefore the step for generating the initial population for the IBEA, as it is given in the description of the IBEA in Sect. 10.1, should be omitted. The description of TS procedure can be found in Sect. 10.2.1.

10.3.1 Cross-Entropy Algorithm

A cross-entropy (CE) procedure, proposed in the PLA2, is perceived as the procedure for preparing some solution basis for further improvement by procedure IBEA. In CE procedure a cross-entropy method first proposed in [12] is used since it was effective in solving various difficult combinatorial optimization problems [13]. Because in CE procedure a solution is viewed as a vector of *n* tasks, we would like to know the probability of locating task J_i on a particular place *j* in the vector. For this reason we introduce two success probability vectors $\hat{p}_j$ and $\hat{p}'_{ji}$ related to each task J_i and its place *j* in solution *S*. Vector $\hat{p}_j = \{p_{ji} \mid 1 \le i \le n\}$, $1 \le j \le n$ contains p_{ji} values, which is the probability that on place *j* there will be located task *i*. Vector $\hat{p}'_{ji} = \{p_{jil} \mid 1 \le l \le W_i\}$, $1 \le j \le n$, $1 \le i \le n$ contains p_{jil} values, which is the probability that on place *j* task *i* will be executed in mode *l*. A procedure CE using cross-entropy method for combinatorial optimization described in [13] and modified for solving problem Θ_Z is shown in the following pseudo code:

```
Procedure CE
Begin
  Set ic:= 1 (ic - iteration counter), ic^stop - maximal
  number of iterations, a:= 1.
```

Set $$\hat{p}_j = \{p_{ji} = 1/n \mid 1 \le i \le n\}, 1 \le j \le n. \quad (10.1)$$

Set $$\hat{p}'_{ji} = \{p_{jil} = 1/W_i \mid 1 \le l \le W_i\}, 1 \le j \le n, 1 \le i \le n. \quad (10.2)$$

While $ic \le ic^{stop}$ do

Generate a sample S_1, S_2, … , S_s, … , S_N of solutions with success probability vectors $\hat{p}_j$ and $\hat{p}'_{ji}$.

Order S_1, S_2, … , S_s, … , S_N by nondecreasing values of their fitness function.

Set $$\gamma = \lceil \rho \cdot N \rceil, \rho \in (0,1). \quad (10.3)$$

Set $$\hat{p}_j = \left\{ p_{ji} = \frac{\sum_{s=1}^{\gamma} I(S_s(j) = i)}{\gamma} \mid 1 \le i \le n \right\}, \quad (10.4)$$

$1 \le j \le n$, $I(S_s(j) = i) = 1$, $I(S_s(j) \ne i) = 0$, where $S_s(j)$ - the number of the task located on j-th place in s-th solution S.

Set $$\hat{p}'_{ji} = \left\{ p_{jil} = \frac{\sum_{s=1}^{\gamma} I(S_s(ji) = l)}{\gamma} \mid 1 \le l \le W_i \right\}, \quad (10.5)$$

$1 \le j \le n$, $1 \le i \le n$, $I(S_s(ji) = l) = 1$, $I(S_s(ji) \ne l) = 0$, where $S_s(ji)$ - an execution mode of task i located on j-th place in s-th solution S.

Save the first $h = \lceil K \cdot PS / ic^{stop} \rceil$ best solutions from the ordered sample into P_0 under address a. Set $a := a + h$.

Set $ic := ic + 1$.

EndWhile.

EndProc_CE.

In the presented pseudo code, a parameter N is the number of solutions in a sample generated in each iteration. A parameter ρ determines the percentage of the best solutions in the current sample that are used to calculate new values for vectors $\hat{p}_j$ and $\hat{p}'_{ji}$. Both parameters were determined empirically and set $N = 1000$ and $\rho = 0{,}2$. Parameters K—the number of islands and PS—the population size are defined in procedure IBEA and PBEA respectively.

10.3.2 Computational Experiment

The considered cross-entropy-based population learning algorithm for solving discrete-continuous scheduling problems with continuous resource discretisation (PLA2), was implemented and tested. Results were compared to the best known obtained by a genetic GAVRdyskr, tabu search, simulated annealing algorithms [7], IBEA [4], and PLA [8] (GAVRdyskr is denoted in the rest of the text as G_{dskr}).

10.3.2.1 Assumptions of the Experiment

For the testing purposes, the following set of assumptions has been formulated and implemented:

- all tasks J_i can be processed in W modes, i.e. $W_i = W$, $i = 1, 2, \ldots, n$, where $W \in \{10, 20, 50\}$.
- continuous resource is discretised uniformly and the amount of the continuous resource assigned to task J_i in mode l_i can be calculated as:

$$u_i^{l_i} = \frac{l_i}{W_i}, \quad l_i = 1, 2, \ldots, W_i, \quad i = 1, 2, \ldots, n \tag{10.6}$$

- task processing rate function f_i is concave and its value can be calculated as:

$$f_i = u_i^{l_i 1/\alpha_i}, \quad \alpha_i \in \{1, 2\}, \quad i = 1, 2, \ldots, n \tag{10.7}$$

- the processing time of task J_i in mode $l_i = 1, 2, \ldots, W_i$ can be calculated as:

$$\tau_i^{l_i} = \frac{\tilde{x}_i}{f_i}, \; i = 1, 2, \ldots, n \tag{10.8}$$

- task sizes $\tilde{x}_i$, $i = 1, 2, \ldots, n$ were generated from interval [1, 1000] with uniform probability distribution,
- the number of tasks $n \in \{10, 20\}$ and the number of machines $m \in \{2, 3\}$.

For the testing purposes three combinations of $n \times m$ were considered—10×2, 10×3, and 20×2. For each considered combination of $n \times m$, 100 instances of problem Θ_Z, have been generated, which together with three discretisation levels (10, 20, 50) makes available 900 instances of the problem. Each instance was tested 24 times. A relative error (RE) of the solutions found by PLA2 with respect to best-known solutions was used to evaluate the quality of PLA2. The value of RE calculated using equation RE = $(Q_{PLA2} - Q_{best\text{-}known})/Q_{best\text{-}known}$ for each instance was used to find average and maximum relative errors.

10.3.2.2 Fine Tuning of PLA2

All three procedures used in the PLA2 have a stochastic nature. The lack of mathematically proved rules makes it difficult to deduce values of the parameters used by the respective procedures. Instead, these values have to be set experimentally. An extensive computational experiment covered all three procedures. In each case we have been looking for the most efficient settings, that is such settings that have been yielding highest quality solutions within some pre-set number of the fitness function evaluations. We tried not only to choose the most proper parameters of the algorithm, but also to find the sequence of the learning stages such that quality of the found solutions was highest possible. Because PLA2 is one more attempt to use an evolutionary approach for coping with the discrete-continuous scheduling problem our goal was to achieve better results for the same number of the fitness function evaluations equal 720000 which was used in the previous trials. While determining the most proper sequence of the learning stages we considered two test versions of PLA2. In these two test versions, the primary solutions were generated according to the rule: random task order in a solution and random task processing mode. In the first test version, primary solutions were improved on the second stage by the IBEA, then on the third stage the best found by the IBEA solution was improved by procedure TS. However, we implemented in PLA2 more efficient second version, in which on the second stage, there was generated only one solution by procedure TS and next added to the set of the primary solutions found on the first stage. This way obtained set of solutions was improved on the third stage by the IBEA. The second version improved 64,67% of 300 best known solutions, while the first—57%, and the average of RE_{avg} was respectively 2,85% and 3,04%. In the final version of the PLA2 we replaced simple first stage procedure, which was used in the PLA for generating the set of the primary solutions, by procedure CE. The percent distribution of the whole number of fitness function evaluations among particular learning procedures is as follows: CE—7%, TS—30%, IBEA—63%. The further tuning concerned the particular procedures engaged in the PLA2.

Procedure CE used to yield the set of primary solutions instead of their random generating increased the percentage of the improved best known solutions from 64,67 to 80,33%, and decreased average of RE_{avg} from 2,85 to 2,09%. For test purposes we designed two versions of procedure CE—with accumulation of the values of success probability vectors $\hat{p}_j$ and $\hat{p}'_{ji}$ in each iteration, and without accumulation. In version with accumulation we calculated the values of vectors $\hat{p}_j$ and $\hat{p}'_{ji}$ in iteration $ic + 1$ using equation:

$$\hat{p}_j(ic+1) = \hat{p}_j(ic-1) + \hat{p}_j(ic) \tag{10.9}$$

$$\hat{p}'_{ji}(ic+1) = \hat{p}'_{ji}(ic-1) + \hat{p}'_{ji}(ic) \tag{10.10}$$

and in the version without accumulation as:

$$\hat{p}_j(ic+1) = \hat{p}_j(ic) \tag{10.11}$$

$$\hat{p}'_{ji}(ic+1) = \hat{p}'_{ji}(ic) \tag{10.12}$$

where values of $\hat{p}_j(ic)$ and $\hat{p}'_{ji}(ic)$ vectors were calculated on the basis of the best γ solutions generated in iteration ic. The main disadvantage of the version with accumulation was sticking in local optima, which explains including the version without accumulation into the final version of the PLA2. In the final implementation of CE procedure we set sample size $N = 1000$ and $\rho = 0{,}2$.

While tuning TS procedure we also considered two versions of it. In both versions of the procedure we generated "a task neighbourhood" set of solutions of a solution S by moving a task $J_i \in S$ from place i to the rest $n - 1$ places in S. In the first version of TS procedure we generated an additional "mode neighbourhood" set of S by assigning to each task $J_i \in S$ one by one in a row the rest all of its $W - 1$ modes, assuming that all tasks can be executed in W modes. The best solution of the iteration was determined as the best across both sets. In the second version of TS procedure, while generating "task neighbourhood" set after each move, we "tuned" the mode only of a single just moved task by assigning to it one by one in a row its remaining $W - 1$ modes. The best solution of the iteration was determined as the best among all solutions generated in such manner. In the final version of the PLA2 we implemented the first TS procedure version as more efficient.

Fine tuning of the IBEA procedure focused on finding the most effective combination of the parameter values. It is known from the literature, that parameters which have a direct impact on the efficiency of the island model are: the number of islands, the size of population on an island [14, 15], migration size [16, 17], migration interval [17, 18], migration policy [16], migration topology [19], and the heterogeneity of the island model [20]. We restricted ourselves to determining on the way of experiment the number of islands K, the best solutions frequency exchange between islands xfq, stop criterion ic^{stop}, and population size on an island PS. The considered parameters were set respectively: $K = 15$, $xfq = 3$, i.e. islands exchanged their best solutions after three populations of individuals were generated on each island, $ic^{stop} = 2000$, i.e. IBEA stopped after 2000 populations had been generated on each island, $PS = 24$ individuals in a population. While tuning IBEA we also considered selection of individuals from the previous population and generating "wild" individuals to the next population. The probability of selection of individuals from previous to the next population depended on the value of the fitness function. In the present PLA2 implementation, tasks' places and tasks' processing modes in a solution vector were determined according to the uniform distribution. Crossover and mutation operators were applied to individuals selected from the previous population. Two individuals took part in each crossover with number of "genes" that were exchanged between parents chosen at random. Mutation of the selected individuals was performed in three ways with probability 0,25, 0,5 and 0,25 respectively. In the first type of mutation we changed at random task processing mode of a chosen at random task. In the second type—chosen at random task J_i was swapped with task J_{i+1}. In the third type of mutation two chosen at random tasks were swapped. All auxiliary random values used in the crossover and mutation operators were acquired according to the uniform probability distribution.

10.3.2.3 Results of the Experiment

RE_{avg} and RE_{max} of the solutions found by the PLA2, PLA, and G_{dskr} are presented in Table 10.4.

The quality of the solutions found by the PLA2 is, on average, better than the quality of the solutions found by the PLA and G_{dskr}. For example, for case 20 × 2, $W = 20$ $RE_{avg} = -0{,}23\%$, which means that the schedule length of all schedules yielded by the PLA2 was, on average, 0,23% shorter than the best-known. For the same case, $RE_{max} = 7{,}23\%$ means that the longest schedule among all schedules yielded by the PLA2 was 7,23% longer than the best-known. In our tests the PLA2 improved 80,33% of 300 the best known solutions for combinations 10 × 2, 10 × 3, and 20 × 2, and reduced average of RE_{avg} compared to the PLA and G_{dskr}. Table 10.5 shows how many percent on average the solutions found by the PLA2 were better than the solutions found by the PLA and G_{dskr}.

Mean time required by the PLA2 and the PLA to find a solution for 10 × 2 on Pentium (R) 4 CPU 3,00 GHz compiled with aid of Borland Delphi Personal v.7.0 was 5 s. The mean time required by G_{dskr}, which was implemented in C++, to find a solution for 10 × 2 on supercomputer Silicon Graphics Power Challenge XL designed in 64-bit SMP (Symmetrical Multi Processing) architecture on 12 RISC MIPS R8000 processors using 1 GB RAM and 20 GB disc memory was 33 s. Such results make the PLA2 quite effective algorithm for solving problem Θ_Z.

Table 10.4 Comparison of REs obtained by the PLA2, PLA and G_{dskr} for problem Θ_Z

Problem size $n \times m \times W$	Relative error (RE)	Discretisation level								
		$W = 10$			$W = 20$			$W = 50$		
		PLA2 (%)	PLA (%)	Gdskr (%)	PLA2 (%)	PLA (%)	Gdskr (%)	PLA2 (%)	PLA (%)	Gdskr (%)
$10 \times 2 \times W$	RE_{avg}	2,75	1,82	7,59	1,53	2,52	9,18	2,25	2,93	10,41
	RE_{max}	9,29	9,39	15,34	5,94	9,00	15,84	8,32	12,21	17,54
$10 \times 3 \times W$	RE_{avg}	2,99	3,76	14,58	2,20	3,32	15,79	2,19	3,96	17,11
	RE_{max}	10,66	11,58	25,18	8,70	12,40	23,15	33,54	14,31	27,52
$20 \times 2 \times W$	RE_{avg}	2,70	2,49	13,25	–0,23	3,11	15,42	2,44	3,42	17,65
	RE_{max}	9,21	9,48	19,40	7,23	9,65	24,02	9,03	9,65	26,87

Table 10.5 The average of the PLA2's RE_{avg} reduction compared to average RE_{avg} of the PLA and G_{dskr} given in percent

Problem size $n \times m \times W$	PLA (%)	G_{dskr} (%)
$10 \times 2 \times W$	0,25	6,89
$10 \times 3 \times W$	1,22	13,36
$20 \times 2 \times W$	1,37	13,80

10.4 PLA3—Population Learning with Differential Evolution Algorithm

The PLA model can be also viewed as an island model, were islands are connected to each other according to some topology and exchange individuals in order to collectively find best possible solution to the problem. We used four learning procedures to design the PLA3: Cross-Entropy (CE), Differential Evolution (DE), Tabu Search (TS), and a Population-Based Evolutionary Algorithm (PBEA) [21]. The PLA3 extends the earlier designed PLA2 [11] with help of the Differential Evolution (DE) method, first proposed in [22]. The PLA3 inherits from the most efficient version of PLA2 (denoted as AX-m in [23]) heterogeneity of the islands, on which diverse learning procedures are realized, random interconnection structure, and solution exchange adjusted to the specificness of the heterogeneous islands. We will use the terms *learning procedure* and an *island* interchangeably in the rest of the text.

We distinguish two categories of island groups—heterogeneous and homogeneous, dependently on the type of the learning procedures carried out on the islands. We refer to the group of islands as heterogeneous, if the learning procedure carried out on at least one island is different from the learning procedures carried out on the rest of the islands in the group. We refer to the group of islands as homogeneous, if the same learning procedure is carried out on each island in the group. In our work, we will refer to a particular island as heterogeneous (Ht), if TS or CE or DE procedure is carried out on it, and we will refer to an island as a homogeneous (Hm), if the PBEA is carried out on it. Because the PLA3 is the extension of its predecessor PLA2, it inherits among others the solution exchange policy. The solution exchange in the PLA3 is carried out among randomly chosen islands and is inherited from AX-m, which, according to [22], is the most efficient version of PLA2. In the PLA3, the assumptions made for the individuals are the same as the assumptions made for the individuals of the IBEA on the population level, see Sect. 10.1. The pseudo code of the proposed PLA3 is given below.

```
Procedure PLA3
Begin
  Set K ≥ 3 - the number of Hm-islands. Assign PBEA pro-
  cedure to all Hm-islands.
  Assign procedures to Ht-islands: TS procedure to Ht1,
  CE procedure to Ht2, DE procedure to Ht3.
  Set xHm - the total amount of solutions in the initial
  population PHm on all Hm-islands.
  Create an initial population PCE of size N ≥ xHm for
  Ht2 using Cross-Entropy procedure (CE).
  Send the best xHm solutions from PCE to PHm.
  Distribute equally individuals from PHm among all
  Hm-islands.
  Create an individual TSI on Ht1. In TSI all tasks Ji
  are to be executed in mode li = 1 (a mode character-
  ized by minimal quantity of additional resource ui^1
  and maximal task processing time τi^1, 1 ≤ i ≤ n).
  Create an initial population PDE of size xDE using
  Cross-Entropy procedure (CE) on Ht3.
  Improve individuals on all islands using assigned to
  them procedures, cyclically exchanging best solutions
  among randomly chosen islands.
  Output the best solution to the problem.
EndProc_PLA3.
```

The solution exchange among the islands occurs after all islands have carried out the preset number of solution evaluations. Generally, the solution exchange is carried out between a pair of islands chosen at random from all available islands. However, there are some exceptions from that rule. In the PLA3 procedure, we distinguish several cases of the solution exchange which are described as follows.

In the case when the solution exchange is carried out between two randomly chosen homogeneous islands Hm_{r1} (PBEA procedure) and Hm_{r2} (PBEA procedure), each island in pair sends to the other one its best current solution.

When the exchange is carried out between Ht_1 island (TS procedure) and Hm_r island, as well as between Ht_1 island and Ht_3 island (DE procedure), each island in pair sends to the other one its best current solution. The exchange procedure between Ht_1 island and Ht_2 island (CE procedure) is described in the next paragraph.

When Ht_2 island (CE procedure) participates in the exchange, the transfer of the solutions is asymmetric. From all available islands Ht_2 receives $\gamma_{CE} = \rho_{CE} \cdot N$ solutions in total, where N is the population size on the Ht_2 island, and $\rho_{CE} = 0{,}2$. In this case of exchange, Ht_2 receives K the best current solutions from all Hm islands, ten copies of the best current solution from Ht_1 island, and $\min(\gamma_{CE} - K - 10, x_{\mathrm{DE}})$ best current solutions from Ht_3 island (DE procedure). If $x_{\mathrm{DE}} < \gamma_{CE} - K - 10$, then Ht_2 additionally receives $\gamma_{CE} - K - 10 - x_{\mathrm{DE}}$ solutions from a randomly chosen island Hm_r and next to it consecutive islands Hm_{r+1}, Hm_{r+2}, …, and Hm_{r+q} until the total number of the solutions received by Ht_2 is equal to γ_{CE}. When $r + q > K$, the numbering of the following consecutive Hm-islands starts with the island number 1. On the other hand, the transfer of the solutions from Ht_2 to the rest of the islands is carried out as follows. When the randomly chosen island in pair is Hm_r, Ht_2 sends its best *PS* current solutions to it, where *PS* is the population size on every Hm-island, defined in PBEA procedure. When the other island in pair is Ht_1, Ht_2 sends its best current solution to it. When the other island in pair is Ht_3, Ht_2 sends its best $\min(\gamma_{CE}, \gamma_{DE-CE})$ current solutions to Ht_3 island, where the value of $\gamma_{DE-CE} = \rho_{DE-CE} \cdot x_{\mathrm{DE}}$, and $\rho_{DE-CE} = 0{,}333$. These solutions substitute the worst solutions on Ht_3 island.

When the exchange is carried out between Ht_3 island (DE procedure) and Hm_r island, they exchange their $\min(PS, \gamma_{DE\text{–}Hm})$ best current solutions, where *PS* is the population size on every Hm-island and $\gamma_{DE\text{–}Hm} = \rho_{DE\text{–}Hm} \cdot x_{\mathrm{DE}}$, $\rho_{DE\text{–}Hm} = 0{,}333$. The solution exchange between Ht_3 and Ht_1, and Ht_3 and Ht_2 islands is described in two previous paragraphs.

10.4.1 Computational Experiment

The proposed population learning algorithm PLA3 for solving discrete-continuous scheduling problems with continuous resource discretisation was implemented and tested. There were 12 islands used in the PLA3 altogether, namely, the number of homogeneous islands was set to $K = 9$ with the PBEA procedure assigned to them and 3 heterogeneous islands. TS procedure was carried out on Ht_1, CE procedure on Ht_2, and DE procedure on Ht_3 island. The size of the population on every Hm-island was set to $PS = 24$. In the TS procedure, the size of the Tabu List (TL) was set to 500 solutions. In the CE procedure, parameters N and ρ_{CE} were set $N = 1000$ and $\rho = 0{,}2$ respectively. The size of the population on Ht_3 island was set $x_{DE} = 2000$, which is different from the sizes considered in [24], where x_{DE} {20, 40, 80, 60, 100}. The rest of the parameters necessary to carry out the differential evolution algorithm were set to the same values as in [24], namely the scale factor A which controls the evolution rate of the population was set $A = 1{,}5$ and the values of the variable *rand* $\in$ [0, 1]. The crossover constants Cr_p and Cr_m which control the probability that the trial individual will receive the actual individual's genes were set $Cr_p = 0{,}2$ and $Cr_m = 0{,}1$, where p and m in the notations Cr_p and Cr_m stand for tasks' positions and modes. For testing purposes three combinations of $n \times m$ were considered (n—the number of tasks and m—the number of machines): 10×2, 10×3, and 20×2. For each combination $n \times m$, 100 instances of a problem Θ_Z were generated and three discretisation levels W were considered: 10, 20, and 50. This way we considered nine sizes of the problem: $10 \times 2 \times 10$, $10 \times 2 \times 20$, $10 \times 2 \times 50$, $10 \times 3 \times 10, \ldots$, $20 \times 2 \times 50$, which makes 900 instances of the problem in total. Each instance was tested 43 times. Mean time required by the PLA3 to find a solution for the problem sizes 10×2 and 10×3 for all discretisation levels on a PC under 64-bit operating system Windows 7 Enterprise with Intel(R) Core(TM) i5-2300 CPU @ 2,80 GHz 3,00 GHz, RAM 4 GB compiled with aid of Borland Turbo Delphi for Win32 was approximately 2–3 s, and for the problem size 20×2 for all discretisation levels approximately 4–6 s.

In order to evaluate the efficiency of the PLA3, we have used three types of relative errors: minimum, average, and maximum relative error of the solutions yielded by the algorithm. Relative errors (REs) of the solutions compared to the best-known solutions were calculated according to the formulae $RE = (Q_{algm} - Q_{best\text{-}known})/Q_{best\text{-}known}$, where Q_{algm}, $Q_{best\text{-}known}$—the schedule length of a solution found by the considered algorithm and the best-known solution respectively. The set of the best-known solutions was determined by the authors while using all designed by them algorithms and procedures for solving problem Θ_Z. We have determined RE_{max} for every size of the considered problem as a maximum RE across 4300 REs calculated while solving 100 instances, run 43 times each. We have also determined RE_{avg} as a mean value of 4300 REs obtained within 43 runs of 100 instances of the considered problem. We have compared the REs of the solutions found by the PLA3 to the REs of the solutions found by AX-m—the most efficient version of the PLA2 described in [23]. The values of RE_{avg} and RE_{max} for the PLA3 and AX-m (PLA2) for all problem sizes are presented in Table 10.6.

The values of REs in Table 10.6 show how much schedules yielded by the PLA3 were longer than the best known schedule for the same case. For example, for the case $10 \times 2 \times 10$ $RE_{avg} = 2{,}70\%$ means that the schedule length of all schedules yielded by the PLA3 was on average 2,70% longer than the best-known. For the same case, $RE_{max} = 10{,}14\%$ means that the longest schedule among all schedules yielded by the PLA3 was 10,14% longer than the best-known.

Table 10.6 The comparison of the relative errors of solutions found by the PLA3 and AX-m (the most efficient version of the PLA2) for the problem Θ_z

Problem size n × m × W	Relative error (RE)	Discretisation level					
		$W = 10$		$W = 20$		$W = 50$	
		PLA3 (%)	AX-m (%)	PLA3 (%)	AX-m (%)	PLA3 (%)	AX-m (%)
10 × 2 × W	RE_{min}	0,01	0,01	**−0,42**	0,00	**−0,49**	0,00
	RE_{avg}	**2,70**	3,54	**1,53**	2,27	**1,81**	2,47
	RE_{max}	**10,14**	12,33	**5,74**	9,14	**9,02**	11,02
10 × 3 × W	RE_{min}	**−0,30**	0,00	**−1,05**	0,00	**−1,63**	0,00
	RE_{avg}	**3,15**	4,68	**1,81**	3,61	**1,66**	3,31
	RE_{max}	**11,76**	16,07	**11,25**	14,71	**12,21**	14,66
20 × 2 × W	RE_{min}	**−0,19**	0,00	**−1,03**	0,00	**−0,14**	0,00
	RE_{avg}	**4,38**	4,76	**1,71**	2,05	4,10	**3,84**
	RE_{max}	**11,77**	11,81	9,22	**9,08**	**11,25**	12,44

As it could be seen in Table 10.6, the values of the considered types of the REs of the solutions found by the PLA3 for the considered problem sizes $n \times m$ and the discretisation levels W, in 23 out of 27 cases were lower than the values of the REs of the solutions found by AX-m version of the PLA2. The value of the RE that is lower than the RE of another considered algorithm is given in bold font. As it could be seen in Table 10.6, it's impossible to determine unequivocally the discretisation level W for which the values of the REs of the found solutions are always the lowest. However, REs yielded by both algorithms for $W = 20$ in a predominant number of cases are the lowest, thus the following relations between the REs can be formulated: REs($W = 20$) < REs($W = 50$) < REs($W = 10$). This might impose the conclusion, that the high discretisation level does not necessarily ensure the lowest values of the REs and the additional research is needed to identify the most appropriate discretisation of the continuous resource.

In addition, we give in Table 10.7 the percentage of the problem instances for which best solutions found by the PLA3 within 43 runs were better (3rd column) or not worse (4th column) than the best-known ones, specified for all of the considered discretisation levels. As a matter of fact, the third column shows the percentage of the problem instances for which the best-known solutions were improved by the PLA3 within 43 runs.

Finally, it should be mentioned, that within 43 runs, for combination 10 × 2, i.e. 10 tasks scheduled on 2 machines, the PLA3 was able to improve 33 out of 100 best known solutions, for combination 10 × 3 – 60 best known solutions, and for combination 20 × 2, the PLA3 improved 30 best known solutions. Altogether,

Table 10.7 The percentage of the problem instances for which best solutions found by the PLA3 within 43 runs were better (3rd column) or not worse (4th column) than the best-known ones specified for all of the considered discretisation levels

Problem size $n \times m \times W$	Discretisation level W	% of the improved the best-known solutions	% of not worse than the best-known solutions
$10 \times 2 \times W$	10	0	0
	20	9	22
	50	26	34
$10 \times 3 \times W$	10	1	2
	20	21	28
	50	48	54
$20 \times 2 \times W$	10	1	1
	20	28	30
	50	2	2

within 43 runs, the PLA3 improved 123 best known solutions, i.e. 41% of 300 instances of the considered problem. It should be also mentioned, that all the conclusions are valid for the particular implementation of the procedures used in the experiments. The values of some parameters of the learning procedures were determined during their tuning and should be verified on the way of exhaustive experiment.

10.5 IBDEA—Island-Based Differential Evolution Algorithm

In an island-based differential evolution algorithm (IBDEA), proposed in [25], two ideas were exploited, namely, the Differential Evolution method, first proposed in [22], and an island model, adopted for evolutionary computation, e.g. [1, 2, 21]. In the IBDEA, the evolutionary process is performed on an archipelago which consists of cooperating with each other autonomous islands. The population on an island consists of two halves of size x_{DE} each. The individuals in the first half—target vectors, are transformed, with help of mutation and crossover operators, into trial vectors which are placed in the second half of the population. The idea of keeping the offspring in the current population was borrowed from [26]. The whole evolutionary process is carried out using differential evolution algorithm (DEA), proposed in [24], which was adapted by the authors for solving DCSPwCRD. In the IBDEA, the islands cooperate with each other, cyclically sending their best solution to one randomly chosen island. The process of evolution stops, when the predefined number of fitness function evaluations is carried out on the archipelago. The best across all islands individual is the final solution, found by the IBDEA to the considered problem. In the rest of the paper, we will use notions "an individual" and "a solution" interchangeably.

In the IBDEA, the assumptions made for the individuals are the same as the assumptions made for the individuals of the IBEA on the population level, see Sect. 10.1. The general description of the proposed IBDEA is given below.

```
Procedure IBDEA
Begin
  Set K ≥ 2 (K - the number of islands). Assign DEA pro-
  cedure to all islands.
  Assume the population of individuals P_k on an island k
  consists of two halves P^1_k and P^2_k, i.e. P_k = P^1_k + P^2_k,
  and |P_k| = x_P = 2·x_DE, |P^1_k| = x_DE, |P^2_k| = x_DE.
  Generate an initial population of individuals P^1_k of
  the size x_DE on every island k, k = 1, 2, … , K.
  Improve individuals on all islands with the DEA proce-
  dure, cyclically exchanging best individuals among
  randomly chosen pairs of islands.
  Stop after n_ev number of fitness function evaluations
  on the archipelago have been carried out.
  Output the best solution to the problem.
EndProc_IBDEA.
```

The individuals in the initial population are generated in such a way, that the position of a task in vector S, as well as the task's processing mode is chosen at random with the uniform distribution.

The solution exchange among the islands occurs cyclically, after $n_{ex} \ll n_{ev}$ number of the fitness function evaluations which have been carried out on every island. The pairs of islands, chosen at random from all the islands, exchange between themselves their best solutions. The random interconnection topology among islands was chosen as the most efficient according to [23].

The DEA procedure used in the IBDEA is described by the following pseudo code.

Procedure DEA

Begin

For each individual (a target vector S_{tg}) in population P^1_k do:

Choose at random from P^1_k three vectors: S_0, S_1, S_2.

Carry out the transformation stage:

create $S' = [S'(p_i) \mid i = 1, 2, \ldots, n]$ for each of S_{tg}, S_0, S_1, S_2, where S' is a tasks' positions vector and $S'(p_i)$ is the position of task i in S,

create modes vector $S''=[S''(l_i) \mid i = 1, 2, \ldots, n]$ for each of S_{tg}, S_0, S_1, S_2, where $S''(l_i)$ is the mode of task i, in which task i is processed in S.

Carry out the mutation stage:

create $M' = [M'(p_i) \mid i = 1, 2, \ldots, n]$ - a tasks' positions mutant vector, calculating $M'(p_i)$ - the mutated position of task i as:

$$M'(p_i) = S'_0(p_i) + A \cdot rand \cdot (S'_1(p_i) - S'_2(p_i)), \quad (10.13)$$

create $M'' = [M''(l_i) \mid i = 1, 2, \ldots, n]$ - a modes' mutant vector, calculating $M''(l_i)$ - the mutated mode of task i as:

$$M''(l_i) = S''_0(l_i) + A \cdot rand \cdot (S''_1(l_i) - S''_2(l_i)). \quad (10.14)$$

Carry out the crossover stage:

create a task trial vector $T' = [T'(p_i) \mid i = 1, 2, \ldots, n]$, determining $T'(p_i)$ - the position of task i in T, as:

$$T'(p_i) = \begin{cases} M'(p_i) \text{ if } rand \le Cr_p \text{ or } i = rand(j) \\ S'_{tg}(p_i) \text{ if } rand > Cr_p \text{ and } i \ne rand(j), \end{cases} \quad (10.15)$$

create a mode trial vector $T'' = [T''(l_i) \mid i = 1, 2, \ldots, n]$, determining $T''(l_i)$ - the mode of task i in T, as:

$$T''(l_i) = \begin{cases} M''(l_i) \text{ if } rand \leq Cr_l \text{ or } i = rand(j) \\ S''_{tg}(l_i) \text{ if } rand > Cr_l \text{ and } i \neq rand(j), \end{cases} \quad (10.16)$$

Carry out the repair stage:

order the tasks in T' according to the ascending values of $T'(p_i)$, this way the new repaired tasks' positions $T'_r(p_i)$ are determined,

create $T'_r = [T'_r(p_i) | i = 1, 2, \ldots, n]$ - a repaired tasks' positions vector,

create $T''_r = [T''_r(l_i) | i = 1, 2, \ldots, n]$ - a repaired tasks modes vector, where the repaired tasks modes $T''_r(l_i)$ are determined as follows:

$$T''_r(l_i) = \begin{cases} 1 \text{ if } T''(l_i) \leq 1, \\ W \text{ if } T''(l_i) \geq W_i, \\ \lfloor T''(l_i) \rfloor \text{ if } 1 < T''(l_i) < W_i. \end{cases} \quad (10.17)$$

End_For

Combine T'_r and T''_r into $T = [c_i | 1 \leq i \leq n]$.

Evaluate T and save in P^2_k.

Select the best x_{DE} individuals from P_k in order to create the next generation of individuals.

Stop after the pre-set number of fitness function evaluations have been carried out.

EndProc_DEA.

A scale factor A, used in DEA procedure, controls the evolution rate of the population. The values of the variable $rand \in [0, 1]$. The crossover constants Cr_p and Cr_l control the probability, that the trial individual will receive the target individual's tasks positions or modes, where p and l in the notations Cr_p and Cr_l stand for tasks positions and modes respectively.

10.5.1 Computational Experiment

Proposed island-based differential evolution algorithm (IBDEA) for solving discrete-continuous scheduling problem with continuous resource discretisation Θ_Z was implemented and tested. There were 19 islands used to realize the IBDEA. The differential evolution algorithm (DEA), described in [24], has been adapted for solving considered Θ_Z and assigned to every island in the IBDEA. After preliminary tuning, the size of the population on every IBDEA island was set $x_{DE} = 200$, which is different from the sizes considered in [24], where $x_{DE} \in \{20, 40, 80, 60, 100\}$. The rest of the parameters necessary to carry out the differential evolution algorithm were set to the same values as in [24], namely the scale factor A, which controls the evolution rate of the population, was set $A = 1{,}5$ and the values of the variable *rand* $\in [0, 1]$. The crossover constants Cr_p and Cr_l which control the probability that trial individual will receive actual individual's tasks or modes were set $Cr_p = 0{,}2$ and $Cr_l = 0{,}1$, where p and l in the notations Cr_p and Cr_l stand for tasks positions and modes respectively. On every IBDEA island, an initial population of feasible individuals was generated using the uniform distribution equal $1/n$ for the tasks, and $1/W$ for the task's modes. For testing purposes three combinations of $n \times m$ were considered (n—the number of tasks and m—the number of machines): 10×2, 10×3, and 20×2. For each combination $n \times m$ 100 instances of a problem Θ_Z were generated and three discretisation levels W were considered: 10, 20, and 50. This way we considered nine sizes of the problem: $10 \times 2 \times 10$, $10 \times 2 \times 20$, $10 \times 2 \times 50$, $10 \times 3 \times 10, \ldots, 20 \times 2 \times 50$, which makes 900 instances of the problem in total. Each instance was tested 43 times. Mean time required by the IBDEA to find a solution for the problem sizes 10×2 and 10×3 for all discretisation levels on a PC under 64-bit operating system Windows 7 Enterprise with Intel(R) Core(TM) i5-2300 CPU @ 2.80 GHz 3.00 GHz, RAM 4 GB compiled with aid of Borland Turbo Delphi for Win32 was approximately 2–3 s, and for the problem size 20×2 for all discretisation levels approximately 5–6 s.

In order to evaluate the efficiency of the IBDEA, we have used three types of relative errors: minimum, average, and maximum relative error of the solutions yielded by the algorithm. Relative errors (REs) of the solutions compared to the best-known solutions were calculated according to the formulae: $\mathrm{RE} = (Q_{algm} - Q_{best\text{-}known})/Q_{best\text{-}known}$, where Q_{algm}, $Q_{best\text{-}known}$—the schedule length of a solution found by the considered algorithm and the best-known solution respectively. The set of the best-known solutions was determined by the authors while using all designed by them algorithms and procedures for solving problem Θ_Z. We have determined $\mathrm{RE}_{\min}$, $\mathrm{RE}_{\mathrm{avg}}$, and $\mathrm{RE}_{\max}$ for every size of the considered problem as a minimum, average, and maximum RE, respectively, across 4300 REs calculated, while solving each of the 100 instances 43 times. We have compared the REs of the solutions found by the IBDEA built on 19 islands to the REs of the solutions found by the IBDEA built on a single island, in other words the DEA itself. The values of such parameters as A, the variable *rand*, Cr_p and Cr_l were set to the same values as in the IBDEA. We have also compared the REs of the solutions found by the IBDEA and the DEA to the REs of the solutions found by the PLA3 described in [21]. The values of RE_{min}, RE_{avg} and RE_{max} for the IBDEA, the DEA and the PLA3 for all problem sizes and considered discretisation levels are presented in Table 10.8. The smallest values of the respective REs for particular cases are given in bold font.

The values of REs in Table 10.8 show how much schedules yielded by the IBDEA were longer than the best known schedule for the same case. For example, for the case 10 × 2 × 10 RE_{avg} = 2,31% means that the schedule length of all schedules yielded by the IBDEA was on average 2,31% longer than the best-known. For the same case, RE_{max} = 7,90% means that the longest schedule among all schedules yielded by the IBDEA was 7,90% longer than the best-known. For the case 10 × 3 × 10 for the IBDEA and the PLA3, RE_{min} = −0,3% is negative, which means that the schedules found by the algorithms were shorter than the best-known for 0,3%. The algorithm whose REs values are the smallest is considered to be more efficient than the others. In Table 10.8, each of three considered algorithms is characterised by nine values of each type of REs for each considered problem size $n \times m \times W$. In 9 out of 9 cases, REs_{max} of the solutions, found by the IBDEA, were smaller than the REs_{max} of the DEA and the PLA3. The remaining REs_{avg} and REs_{min} of the IBDEA were the smallest in 6 and in 2 cases respectively. However, REs_{avg} and REs_{min} of the PLA3, in 3 and in 4 cases out of 9 were the smallest. In all cases, all types of the DEA's REs were the largest. As a general conclusion, we point out that in 17 cases out of 27, the IBDEA shows the smallest REs, and no other algorithm shows the same or better results. The PLA3 was the best in 8 cases out of 27, and the DEA only once achieved the same result as the IBDEA and the PLA3 (RE_{min} for the problem size 10 × 2 × 10).

As it could be seen in Table 10.8, it's impossible to determine unequivocally the discretisation level W for which the values of the REs of the found solutions are always the smallest. However, REs yielded by all considered algorithms for $W = 20$ in a majority of cases were the smallest, thus the following relations between the REs can be formulated: $REs(W = 20) < REs(W = 50) < REs(W = 10)$. This might impose the conclusion, that high discretisation level does not necessarily ensure the smallest values of the REs and the additional research is needed to identify the most appropriate discretisation of the continuous resource.

The computational experiment shows, that the island model exploiting DE finds solutions whose relative errors (REs) are smaller than the REs of the solutions found by the DEA alone, see Table 10.8. The above statement is true under assumption, that on every island of the IBDEA operates the DEA, and that the size of the population on every island is the same as the size of the population in the DEA. In major number of cases, RE_{min} and RE_{avg} of the solutions found by the IBDEA were smaller than the RE_{min} and RE_{avg} of solutions found by the DEA and the PLA3, which also exploits the island model. The direct benefit of the proposed algorithm is its ability to find high quality solutions with a smaller dispersion of RE's values. In our experiment, in all cases, RE_{max} of the solutions found by the IBDEA were the smallest. The promising results achieved by the IBDEA might

Table 10.8 The comparison of the relative errors RE of solutions found by the IBDEA, the DEA, and the PLA3 for problem Θ_z

Problem size $n \times m \times W$	Relative error (RE)	Discretisation Level								
		$W = 10$			$W = 20$			$W = 50$		
		IBDEA (%)	DEA (%)	PLA3 (%)	IBDEA (%)	DEA (%)	PLA3 (%)	IBDEA (%)	DEA (%)	PLA3 (%)
$10 \times 2 \times W$	RE_{min}	**0,01**	**0,01**	**0,01**	−0,33	0,00	**−0,42**	**−0,65**	−0,53	−0,49
	RE_{avg}	**2,31**	3,86	2,70	**0,93**	2,86	1,53	**0,83**	3,22	1,81
	RE_{max}	**7,90**	11,52	10,14	**4,57**	10,72	5,74	**4,91**	11,86	9,02
$10 \times 3 \times W$	RE_{min}	**−0,30**	−0,10	**−0,30**	−0,75	−0,04	**−1,05**	**−1,71**	−1,12	−1,63
	RE_{avg}	**2,60**	4,29	3,15	**0,96**	3,18	1,81	**0,46**	3,02	1,66
	RE_{max}	**10,04**	13,31	11,76	**8,60**	17,16	11,25	**7,63**	16,80	12,21
$20 \times 2 \times W$	RE_{min}	0,78	0,33	**−0,19**	−0,78	−0,20	**−1,03**	1,29	−0,03	**−0,14**
	RE_{avg}	5,03	4,93	**4,38**	5,22	4,18	**1,71**	6,94	5,24	**4,10**
	RE_{max}	**10,11**	11,91	11,77	**9,20**	11,37	9,22	**10,53**	13,22	11,25

suggest its superiority over the DEA, however in order to make the final conclusion more extensive research is needed.

Finally, it should be also mentioned, that all the conclusions are valid for the particular implementation of the procedures used in the experiments. The values of some parameters of the learning procedures were determined during their preliminary tuning and should be verified on the way of the exhaustive experiment.

References

1. Alba, E., Troya, J.: Analysis of synchronous and asynchronous parallel distributed genetic algorithms with structured and panmictic islands. In: Rolim, J. et al., (eds.) Proceedings of the 10th Symposium on Parallel and Distributed Processing, pp. 248–256. San Juan, Puerto Rico, USA, 12–16 April (1999)
2. Belding, T.C.: The distributed genetic algorithm revisited. In: Eshelman, L.J. (ed.) Proceedings of the Sixth International Conference on Genetic Algorithms, pp. 114–121. Morgan Kaufmann, San Francisco CA (1995)
3. Gordon, V.S., Whitley, D.: Serial and parallel genetic algorithms as function optimizers. In: Forrest, S. (ed.) Proceedings of the Fifth International Conference on Genetic Algorithms, pp. 177–183. Morgan Kaufmann, San Mateo, CA (1993)
4. Czarnowski, I., Gutjahr, W.J., Jędrzejowicz, P., Ratajczak, E., Skakowski, A., Wierzbowska, I.: Scheduling multiprocessor tasks in presence of correlated failures. In: Luptaćik, M., Wildburger, U.L. (eds.) Central European Journal of Operations Research, vol. 11, iss. 2, pp. 163–182. Physika-Verlag, Springer, Heidelberg (2003)
5. Jędrzejowicz, P., Skakovski, A., Czarnowski, I., Szreder, H.: Evolution-based scheduling of multiple variant and multiple processor programs. In: Hertzberger, L.O., Sloot P.M.A. (eds.) Future Generation Computer Systems, vol. 17, pp. 405–414. Elsevier, The Netherlands (2001)
6. Jędrzejowicz, P., Skakovski, A.: An island-based evolution algorithm for discrete-continuous scheduling with continuous resource discretisation. In: Proceedings of the 2nd IEEE International Conference on Computational Cybernetics ICCC 2004, 30 Aug–1 Sep 2004. Vienna University of Technology, Austria (2004)
7. Różycki, R.: Zastosowanie algorytmu genetycznego do rozwiązywania dyskretno-ciągłych problemów szeregowania. PhD Dissertation, Poznań University of Technology, Poland (2000)
8. Jędrzejowicz, P., Skakovski, A.: A population learning algorithm for discrete-continuous scheduling with continuous resource discretisation. In: Chen, Y., Abraham, A. (eds.) Proceedings of 6th International Conference on Intelligent Systems Design and Applications (ISDA 2006), vol. 2, spec. sess.: Nature Imitation Methods Theory and practice (NIM' 06), pp. 1153–1158. Jinan, Peoples R. of China (2006)
9. Jędrzejowicz, P.: Social learning algorithm as a tool for solving some difficult scheduling problems. Found. Comput. Decis. Sci. **24**, 51–66 (1999)
10. Glover, F.: Tabu search: a tutorial. Interfaces **20**, 74–94 (1990)
11. Jędrzejowicz, P., Skakovski, A.: A cross-entropy based population learning algorithm for discrete-continuous scheduling with continuous resource discretisation. Neurocomputing **73** (4–6), Special Issue: SI:655–660 (2010)
12. Rubinstein, R.Y.: Optimization of computer simulation models with rare events. Eur. J. Op. Res. **99**, 89–112 (1997)
13. De Boer, P.-T., Kroese, D.P., Mannor, S., Rubinstein, R.Y.: A tutorial on the cross-entropy method. Ann. Op. Res. **134**(1), 19–67 (2005)

14. Cantu-Paz, E., Goldberg, D.E.: Are multiple runs of genetic algorithms better than one?. In: Proceedings of the Genetic and Evolutionary Computation Conference (2003)
15. Whitley, D., Rana, S., Heckendorn, R.B.: The island model genetic algorithm: on separability, population size and convergence. J. Comput. Inf. Technol. **7**(1), 33–47 (1999)
16. Cantu-Paz, E.: Migration policies, selection pressure, and parallel evolutionary algorithms. J. Heuristics **7**(4), 31–334 (2001)
17. Skolicki, Z., Kenneth, D.J.: The influence of migration sizes and intervals on island models. In: Proceedings of GECCO' 05, pp. 1295–1302. Washington, DC, USA, 25–29 June (2005)
18. Krink, T., Mayoh, B.H., Michalewicz, Z.: A PACHWORK model for evolutionary algorithms with structured and variable size populations. In: Morgan, K., Banzhaf, W., Daida, J., Eiben, A.E., Garzon, M.H., Honavar, V., Jakiela, M., Smith, R.E. (eds.) Proceedings of the Genetic and Evolutionary Computation Conference, vol. 2, pp. 1321–1328. Orlando, Florida, USA (1999)
19. Sekaj, I.: Robust parallel genetic algorithms with re-initialisation. In: Proceedings of Parallel Problem Solving from Nature—PPSN VIII, 8th International Conference, vol. 3242, pp. 411–419. LNCS, Springer, Birmingham, UK, 18–22 Sep (2004)
20. Skolicki, Z.: An analysis of island models in evolutionary computation. In: Proceedings of GECCO' 05, pp. 386–389. Washington, DC, USA, 25–29 June (2005)
21. Jędrzejowicz, P., Skakovski, A.: Population learning with differential evolution for the discrete-continuous scheduling with continuous resource discretisation. In: IEEE International Conference on Cybernetics (CYBCONF) pp. 92–97. Lausanne, Switzerland, 13–15 June (2013)
22. Storn, R., Price, K.: Differential evolution—a simple and efficient heuristic for global optimization over continuous spaces. J. Global Opt. **11**, 341–359 (1997)
23. Jędrzejowicz, P., Skakovski, A.: Structure vs. efficiency of the cross-entropy based population learning algorithm for discrete-continuous scheduling with continuous resource discretisation. In: Czarnowski, I., Jędrzejowicz, P., Kacprzyk, J. (eds.) Studies in Computational Intelligence. Agent-Based Optimization, vol. 456, pp. 77–102 (2013)
24. Damak, N., Jarboui, B., Siarry, P., Loukil, T.: Differential evolution for solving multi-mode resource-constrained project scheduling problems. Comput. Op. Res. **36**(9), 2653–2659 (2009)
25. Jędrzejowicz, P., Skakovski, A.: Island-based differential evolution algorithm for the discrete-continuous scheduling with continuous resource discretisation. Procedia Comput. Sci. **35**, 111–117 (2014)
26. Kazemipoor, H., Tavakkoli-Moghaddam, R., Shahnazari-Shahrezaei, P.: Differential evolution and simulated annealing algorithms for a multi-skilled project scheduling problem. Am. J. Sci. Res. **33**, 136–146 (2011)

Chapter 11
Performance Evaluation of the Proposed Algorithms

In this section, a research on the performance evaluation and ways of the performance improvement for some of the proposed algorithms are discussed. Some useful properties of the DEA, used in the PLA3 and the IBDEA, and a policy for its performance improvement are also presented.

In Sect. 11.1, the Friedman test carried out to evaluate the performance of 6 metaheuristics: the IBEA, the PLA, the PLA2, the PLA3, the IBDEA, and the DEA, discussed in the previous section, with the view of determining the most efficient of the algorithms is described.

In Sect. 11.2, the influence of the structure of PLA2 on its efficiency was studied. The main goal of the research was to find out to what extent such factors as the interconnection topology of learning procedures (islands) and the way the islands interact with each other and exchange solutions determine the efficiency of the algorithm. For this reason several versions of the PLA2 differing from each other by their structure and migration scheme, were proposed. The research originates from [1].

In Sect. 11.3, we examine the properties of two models of the DE search: a model based on a single population (the implementation of which is denoted as DEA), and a model based on multiple populations, known as the island model. We investigated how the effectiveness of the models under consideration depends on such parameters as the size of a single population x_P, and in the case of the island model, also the number of islands K and the migration rate ex. The research originates from [2].

In Sect. 11.4, we conduct a research on preserving the diversity of a population evolved by a differential evolution algorithm. We designed a decloning procedure which cyclically replaces genetically identical individuals (clones) with randomly generated ones, the detailed description of the decloning procedure is given in Sect. 11.4.2. The goal of our research was to investigate the extent to which performance of the considered differential evolution algorithm depends on such parameters as the population diversification rate, the size of the population, and the

E. Ratajczak-Ropel and A. Skakovski, *Population-Based Approaches to the Resource-Constrained and Discrete-Continuous Scheduling*,
Studies in Systems, Decision and Control 108, DOI 10.1007/978-3-319-62893-6_11

number of fitness function evaluations carried out by the algorithm to yield a solution to the problem. The research originates from [3].

In all experiments, the discrete-continuous scheduling problem with continuous resource discretisation (DCSPwCRD) described in Sect. 9.2.2 was used as a test problem.

11.1 Friedman Test

In this section we describe the non-parametric statistical Friedman test which was carried out to evaluate the performance of 6 metaheuristics: the IBEA, the PLA, the PLA2, the PLA3, the IBDEA, and the DEA, described in the previous chapter, with the view of determining the most efficient of the algorithms. The evaluation is based on ranks assigned to each metaheuristic participating in the experiment. To assign the ranks, a 6 point scale was used, with 6 points for the best and 1 point for the worst result found by the metaheuristics for a particular problem instance. When the results were identical (i.e. there were ties among ranks), the same amount of points equaled the mean of consecutive ranks was assigned to each such result. The metaheuristics were tested on 300 instances of the discretised discrete-continuous scheduling problem Θ_Z generated for three sizes $m \times n$ (n—the number of tasks and m—the number of machines): 10×2, 10×3, and 20×2. The level of the continuous resource discretisation was set at 20, which was assumed as the most beneficial according to the results given in Table 11.20 in Sect. 11.2.1. All considered algorithms carried out 720,000 fitness function evaluations to find a solution with the parameter settings described in the respective sections of Chap. 10. The population size of the DEA was set at 2000. The test aimed at deciding among the following hypotheses:

- H_0—a zero hypothesis: the metaheuristics are statistically equally effective, regardless of the problem instance being solved.
- H_1—an alternative hypothesis: not all metaheuristics are equally effective.

The analysis has been carried out for the significance level $\alpha = 0{,}05$ and 5 degrees of freedom ($df = k - 1 = 6 - 1 = 5$). The respective value of χ^2 statistics calculated for the considered case of 6 metaheuristics and 300 problem instances equals 690,75. The critical value of χ^2 distribution for the assumed values of significance level α and degrees of freedom df equals 11,07. Since the obtained value of the Friedman statistics χ^2 is greater than the critical one, hypothesis H_0 is rejected. Thus, the obtained result proves hypothesis H_1 which claims that not all metaheuristics are equally effective, regardless of problem instances being solved. The means of ranks, obtained for the metaheuristics under the test, are shown in Fig. 11.1.

Additionally, the Kendall's coefficient of concordance (Kendall's W) was calculated. The Kendall's W was used for evaluating the degree of agreement among the obtained assessments. Kendall's W ranges from 0 (no agreement) to 1 (complete

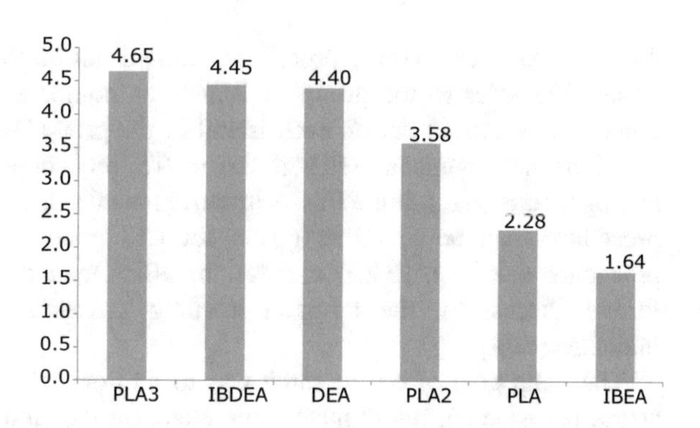

Fig. 11.1 The means of ranks obtained for the metaheuristics under the test

agreement). The intermediate values of W indicate a greater or lesser degree of unanimity among the various assessments. The value of $W = 0{,}4605$, calculated for the results obtained in our Friedman test, indicates a nearly medial degree of unanimity among the assessments. Although, this value of Kendall's W does not allow a reliable conclusion about the superiority of some over others metaheuristics, however the mean values of ranks, calculated for the algorithms, might suggest some ranking with respect to their efficiency. Thus, according to the means of ranks obtained in our test and given in parenthesis, one could rank the metaheuristics from the most to the least efficient as follows: the PLA3 (4,65), the IBDEA (4,45), the DEA (4,4), the PLA2 (3,58), the PLA (2,28), the IBEA (1,64), see also Fig. 11.1. As it could be seen from the ranking proposed and Fig. 11.1, three metaheuristics the PLA3, the IBDEA, and the DEA with similar rank means (4,65, 4,45, 4,40) can be considered as superior to the PLA2, the PLA, and the IBEA with smaller rank means 3,58, 2,28, and 1,64 respectively.

11.2 Structure Versus Efficiency of the Cross-Entropy-Based Population Learning Algorithm (PLA2)

Because most of the algorithms proposed in this work are based on the island model, it was of crucial importance to establish the extent to which the structure and migration scheme of the algorithms determine their performance. These factors were investigated in [1] using the PLA2 as a test algorithm, since it exploits the island model.

The PLA2 model introduced in [4] and described in Sect. 10.3 is viewed in the research as an island model, were islands are connected to each other according to some topology and exchange individuals in order to collectively find best possible solution to the problem. In the PLA2, two categories of island groups are distinguished—heterogeneous and homogeneous, dependently on the type of the learning procedures that are carried out on the islands. We refer to the group of islands as heterogeneous, if the learning procedure carried out on at least one island is different from the learning procedures carried out on the rest of the islands in the group. We refer to the group of islands as homogeneous, if the same learning procedure is carried out on each island in the group. We will refer to a particular island as heterogeneous (Ht), if CE or TS procedure is carried out on it, and homogeneous (Hm), if the PBEA is carried out on it. The description of mentioned procedures can be found as given: for CE procedure see Sect. 10.3.1, for TS procedure see Sect. 10.2.1, and for the IBEA and the PBEA see Sect. 10.1. In further discussion, the terms a learning procedure and an island are used interchangeably.

The main goal of our research was to find out whether a topology of learning stages (or islands), might have some effect on the algorithm's efficiency. For this reason we proposed several versions of the PLA2 that differ from each other by their structure and migration scheme. We will refer to these versions of the PLA2 as algorithms, and some letter code will be assigned to each of them. In order to distinguish the algorithms, we considered two their basic types, each of them having two topology schemes. In the first basic type, all islands participate in the solution evolution and migration at least once. However, some *selected* islands additionally take part in the cyclic solution migration among islands (the letter code for this version will contain letter "S"). In the second basic type—all islands take part in the cyclic solution migration among the islands (the letter code for this version will contain letter "A"). As it was mentioned above, each basic type appears in two topology schemes. In the first topology scheme, islands are located on a directed ring and the individuals migrate among the islands along the ring (the letter code for this scheme will contain letter "O" and we will refer in the rest of the text to this topology as a *ring topology*). In the second topology scheme—individuals migrate between randomly chosen pairs of islands (the letter code for this scheme will contain letter "X" and we will refer to this topology as a *random topology*). Moreover, the letter code for the algorithms in which CE procedure sends multiple solutions to the island-in-pair during the migration phase will contain letter "m". For the version where CE procedure sends a single solution to the island-in-pair during the migration phase the letter code will contain letter "s". Therefore, the letter code "AO-m" stands for the algorithm in which all islands comprise a directed ring of heterogeneous islands and procedure CE sends multiple solutions to the island-in-pair during the migration phase. In this research, six versions of the PLA2 were considered, namely: SO (S-only selected islands take part in the cyclic migration, O—ring topology), SX (S—only selected islands take part in the cyclic migration, X—random topology), AO-m (A—all islands take part in the cyclic migration, O—ring topology, the CE sends multiple solutions), AO-s (A—all islands take part in the cyclic migration, O-ring topology, the CE sends a single solution), AX-m (A—all islands take part in the cyclic migration, X—random topology, CE sends multiple solutions), and AX-s (A—all islands take part in the cyclic migration, X—random topology, the CE sends a single solution). Because all proposed algorithms are versions of the PLA2, they have common bases, which are shown as generalized S-type and A-type algorithms. The pseudo codes, as well as figures illustrating all proposed algorithms are given below.

S-algorithm

Begin

Create an initial population P_0 of the size x_0-1 using procedure cross-entropy (CE).

Create an individual *TSI* in which all tasks J_i are to be executed in mode $l_i = 1$ (a mode characterized by minimal quantity of additional resource u_i^1 and maximal task processing time τ_i^1, $1 \leq i \leq n$).

Improve the individual *TSI* with the tabu search (TS) procedure.

Create population $P_1 = P_0 + TSI$.

Distribute equally individuals from P_1 among all Hm-islands.

Carry out the appropriate Learning stage SO or SX designed for SO and SX algorithms respectively.

Output the best solution to the problem.

End_S-algorithm.

Learning stage SO

Begin

Improve individuals on Hm-islands with procedure IBEA.

End_stage_SO.

Learning stage SX

Begin

Improve individuals on each Hm-island with procedure PBEA, cyclically exchanging best solutions between randomly chosen pairs of Hm-islands.

End_stage_SX.

```
A-algorithm
Begin
  Create an initial population P0 of the size x0 using
  cross-entropy procedure (CE).
  Distribute equally individuals from P0 among all
  Hm-islands.
  Create an individual TSI in which all tasks Ji are to
  be executed in mode li = 1 (a mode characterized by
  minimal quantity of additional resource ui^1 and maxi-
  mal task processing time τi^1, 1 ≤ i ≤ n).
  Send TSI to the Tabu Search (TS) procedure.
  Carry out the appropriate Learning stage AO or AX de-
  signed for AO and AX algorithms respectively.
  Output the best solution to the problem.
End_A-algorithm.
Learning stage AO
Begin
  Create a directed ring of all available islands as
  follows: Hm1, Ht1(TS), Hm2, Hm3, Ht2(CE), … , HmK, where
  K ≥ 3.
  Improve individuals on the islands with the assigned
  to them procedures cyclically sending best solution
  from each island along the ring.
End_stage_AO.
Learning stage AX
Begin
  Improve individuals on all available islands with the
  assigned to the islands procedures cyclically ex-
  changing best solution between randomly chosen pairs
  of islands.
End_stage_AX.
```

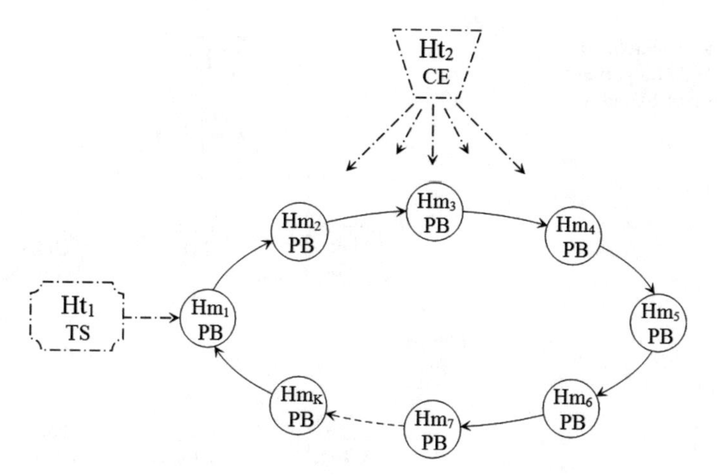

Fig. 11.2 A simplified scheme of SO algorithm based on ring topology

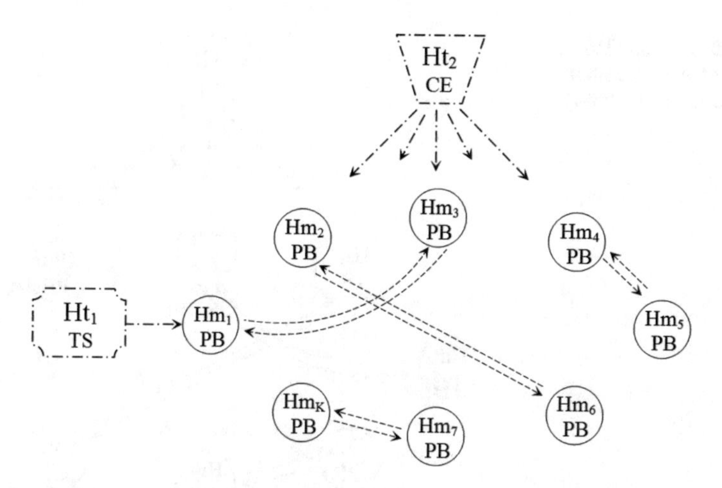

Fig. 11.3 A simplified scheme of SX algorithm based on random topology

In all proposed algorithms, $x_0 = K \cdot PS$, where K—the number of homogeneous islands and PS is the population size on an island defined in procedure IBEA.

In a simplified graphic illustration of the algorithms in Figs. 11.2, 11.3, 11.4 and 11.5, solid lines show islands participating in the cyclic solution migration and dash-dot lines show islands where learning procedures are carried out only once. In all figures, an abbreviation PB denotes the PBEA procedure.

In AO algorithm, CE procedure receives multiple solutions from the homogeneous islands Hm_1, Hm_2 and Hm_3, and sends a single solution to Hm_4 (or Hm_1, when $K = 3$) in its AO-s version, or multiple solutions in its AO-m version. In AX algorithm CE procedure also receives multiple solutions from Hm_1, Hm_2 and Hm_3, and sends to the randomly chosen island a single solution in AX-s algorithm,

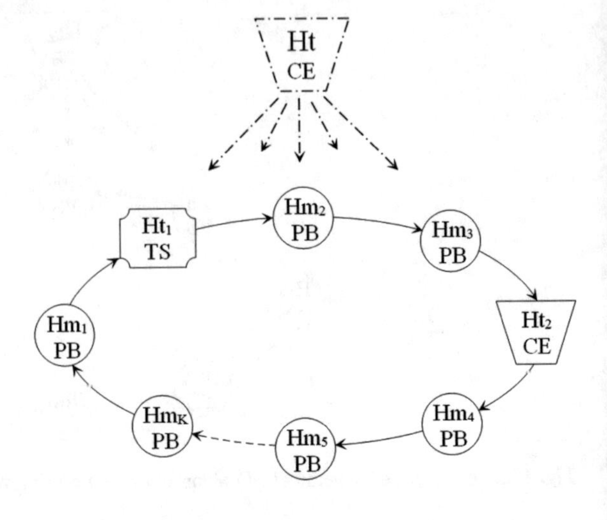

Fig. 11.4 A simplified scheme of AO algorithm based on ring topology

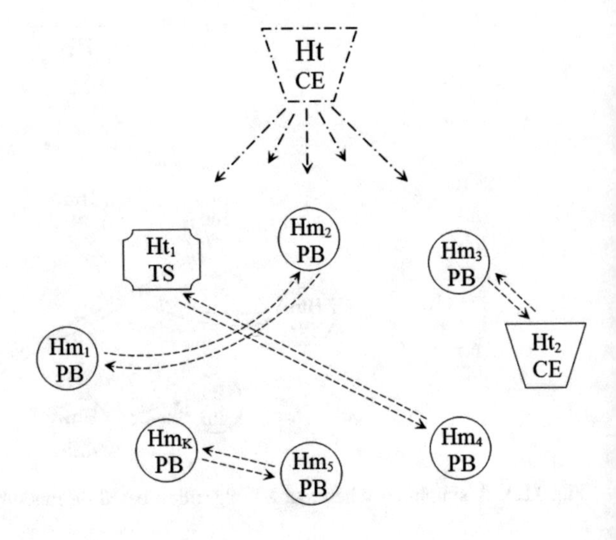

Fig. 11.5 A simplified scheme of AX algorithm based on random topology

or multiple solutions in AX-m algorithm. On all homogeneous islands Hm_i, $i = 1, 2, \ldots, K$ we used the PBEA procedure.

11.2.1 Computational Experiment

All six proposed versions of the cross-entropy based population learning algorithm for solving discrete-continuous scheduling problems with continuous resource discretisation were implemented and tested. The efficiency of all six algorithms was compared to each other, as well as to the tabu search (TS) procedure, used within each of the algorithms which was run in addition as an independent algorithm. In the procedure CE, as it was mentioned earlier, parameters ρ and N were determined empirically and set $N = 1000$ and $\rho = 0{,}2$. The size of the Tabu List (TL) was determined empirically as well, and set to 500 solutions. For testing purposes three combinations of $n \times m$ were considered (n—the number of tasks and m—the number of machines): 10×2, 10×3, and 20×2. For each combination $n \times m$ 100 instances of problem Θ_Z were generated and three discretisation levels W were considered: 10, 20, and 50. This way we considered nine sizes of the problem: $10 \times 2 \times 10$, $10 \times 2 \times 20$, $10 \times 2 \times 50$, $10 \times 3 \times 10$, … , $20 \times 2 \times 50$, which makes 900 instances of the problem in total. In all problem instances, we have used the same as in [5] task processing rate functions already described in Sect. 10.3.2.1.

Each of the considered algorithms carried out about 720000 fitness function evaluations to yield one solution for the instance of the problem. Each instance was tested 43 times by all the proposed algorithms. Mean time required by the considered algorithms to find a solution for the problem sizes 10×2 and 10×3 for all discretisation levels on Pentium (R) 4 CPU 3.00 GHz compiled with aid of Borland Delphi Personal v.7.0 was approximately 4–7 s, and for the problem size 20×2 for all discretisation levels approximately 8–13 s.

In order to evaluate the efficiency of the proposed algorithms we used such parameters as relative errors (minimum, average, maximum) of the solutions yielded by the algorithms, as well as percentage of solutions of the same quality as the best-known solutions. Relative errors (RE) of the solutions compared to the best-known solutions were calculated according to the formulae $RE = (Q_{PLA2} - Q_{best\text{-}known})/Q_{best\text{-}known}$, where Q—the quality of a considered solution. The set of the best-known solutions was determined by the authors while using all designed by them procedures and algorithms, namely the PBEA, the IBEA, TS, the PLA, the PLA2, AX-m, AX-s, SX, SO, AO-m, and AO-s for solving problem Θ_Z. We have determined RE_{min} and RE_{max} for every size of the considered problem as a minimum or respectively maximum RE across 4300 REs calculated while solving each of the 100 instances 43 times. We have also determined RE_{avg} as a mean value of 4300 REs obtained within 43 runs of 100 instances of the considered problem. The values of RE_{min}, RE_{avg} and RE_{max} of the solutions found by all proposed algorithms for all problem sizes are presented in Tables 11.1, 11.2, 11.3, 11.4, 11.5, 11.6, 11.7, 11.8, and 11.9. The values of REs in Tables 11.1, 11.2, 11.3, 11.4, 11.5, 11.6, 11.7, 11.8, and 11.9 show how much schedules yielded by the proposed algorithms were longer than the best known schedule for the same case. For example, in Table 11.1 for the case $10 \times 2 \times 10$ for SO algorithm, $RE_{avg} = 3.28\%$ means that the schedule length of all schedules yielded by SO algorithm was on average 3.28% longer than the best-known. For the same case, $RE_{max} = 9{,}76\%$ means that the longest schedule among all schedules yielded by SO algorithm was 9,76% longer than the best-known. To make it easier to evaluate their efficiency, in Tables 11.1,11.2, 11.3, 11.4, 11.5, 11.6, 11.7, 11.8, and 11.9 below, we have ordered the algorithms according to their REs non-decreasingly.

Table 11.1 The algorithms ordered non-decreasingly according to their RE_{min}, RE_{avg} and RE_{max} for the size 10 × 2 × 10 of the problem Θ_z

Nr	Alg-m	RE_{min} (%)	Alg-m	RE_{avg} (%)	Alg-m	RE_{max} (%)
1	SX	0,01	SO	3,28	SO	9,76
2	AO-m	0,01	SX	3,31	TS	9,91
3	AO-s	0,01	TS	3,49	SX	10,00
4	AX-m	0,01	AX-m	3,54	AX-s	11,39
5	AX-s	0,01	AX-s	3,58	AX-m	12,33
6	TS	0,01	AO-m	6,30	AO-m	14,68
7	SO	0,19	AO-s	6,30	AO-s	16,53

Table 11.2 The algorithms ordered non-decreasingly according to their RE_{min}, RE_{avg} and RE_{max} for the size 10 × 3 × 10 of the problem Θ_z

Nr	Alg-m	RE_{min} (%)	Alg-m	RE_{avg} (%)	Alg-m	RE_{max} (%)
1	AX-s	0,00	SO	4,67	SO	15,92
2	SO	0,00	AX-m	4,68	AX-m	16,07
3	AX-m	0,00	SX	4,69	TS	17,72
4	SX	0,01	AX-s	4,74	SX	18,39
5	TS	0,07	TS	5,36	AX-s	19,33
6	AO-m	0,10	AO-m	8,66	AO-s	25,67
7	AO-s	0,28	AO-s	8,71	AO-m	26,06

Table 11.3 The algorithms ordered non-decreasingly according to their RE_{min}, RE_{avg} and RE_{max} for the size 20 × 2 × 10 of the problem Θ_z

Nr	Alg-m	RE_{min} (%)	Alg-m	RE_{avg} (%)	Alg-m	RE_{max} (%)
1	AX-m	0,00	AX-m	4,76	SO	11,47
2	AX-s	0,26	AX-s	4,86	SX	11,74
3	SX	0,40	SO	5,46	AX-m	11,81
4	SO	0,51	SX	5,56	TS	12,09
5	TS	0,68	TS	6,26	AX-s	12,54
6	AO-m	1,12	AO-s	9,19	AO-m	16,71
7	AO-s	1,16	AO-m	9,31	AO-s	18,19

For the problem size 10 × 2 × 10, according to the values of the RE_{min} in Table 11.1, SO algorithm has the largest RE_{min}, however quite close to the REs of the other algorithms. On the other hand, considering RE_{avg}, SO algorithm has the lowest RE_{avg}, and this way, is the leader in a group of the algorithms: SO, SX, TS, AX-m, AX-s, with similar RE_{avg} values.

Algorithms AO-m, AO-s make another group of the same RE_{avg} values, where RE_{avg} is about twice higher than in the first group. Considering RE_{max}, it is also possible to classify the algorithms into two groups: with low RE_{max}: SO, TS, SX, and with high RE_{max}: AX-s, AX-m, AO-m, AO-s. In the first group the algorithms

Table 11.4 The algorithms ordered non-decreasingly according to their RE_{min}, RE_{avg} and RE_{max} for the size 10 × 2 × 20 of the problem Θ_z

Nr	Alg-m	RE_{min} (%)	Alg-m	RE_{avg} (%)	Alg-m	RE_{max} (%)
1	SX	0,00	SO	2,05	SO	6,67
2	AO-m	0,00	SX	2,07	SX	7,04
3	AO-s	0,00	AX-m	2,27	TS	7,66
4	AX-m	0,00	AX-s	2,33	AX-m	9,14
5	AX-s	0,00	TS	2,36	AX-s	12,40
6	TS	0,00	AO-m	4,99	AO-m	17,18
7	SO	0,00	AO-s	5,01	AO-s	17,60

Table 11.5 The algorithms ordered non-decreasingly according to their RE_{min}, RE_{avg} and RE_{max} for the size 10 × 3 × 20 of the problem Θ_z

Nr	Alg-m	RE_{min} (%)	Alg-m	RE_{avg} (%)	Alg-m	RE_{max} (%)
1	AX-m	0,00	AX-m	3,61	AX-m	14,71
2	SX	0,00	AX-s	3,73	SO	15,18
3	AX-s	0,00	SO	3,87	AX-s	15,89
4	TS	0,00	SX	4,06	SX	16,66
5	SO	0,00	TS	4,95	TS	18,28
6	AO-m	0,00	AO-s	8,13	AO-m	24,24
7	AO-s	0,08	AO-m	8,13	AO-s	26,03

Table 11.6 The algorithms ordered non-decreasingly according to their RE_{min}, RE_{avg} and RE_{max} for the size 20 × 2 × 20 of the problem Θ_z

Nr	Alg-m	RE_{min} (%)	Alg-m	RE_{avg} (%)	Alg-m	RE_{max} (%)
1	SX	0,00	AX-m	2,05	AX-m	9,08
2	AX-m	0,00	SO	2,43	SO	9,76
3	AX-s	0,00	AX-s	2,53	AX-s	9,90
4	SO	0,00	AO-m	2,60	TS	10,81
5	AO-m	0,00	AO-s	2,72	SX	11,00
6	AO-s	0,00	SX	2,84	AO-m	15,19
7	TS	0,53	TS	5,09	AO-s	15,61

differ from 9,76% to 10,00%, while in the second—from 11,39% to 16,53%. For the problem size 10 × 2 × 10, $RE_{min} \in [0{,}01\%, 0{,}19\%]$, $RE_{avg} \in [3{,}28\%, 6{,}30\%]$, $RE_{max} \in [9{,}76\%, 16{,}53\%]$. Generally, according to Table 11.1, the algorithms exploiting the directed ring migration scheme (the ring topology) or random migration scheme (the random topology), built exclusively on homogeneous islands, as well as scheme built on all islands with the random topology perform better, than the algorithms exploiting the ring topology built on all islands—both homogeneous and heterogeneous. This way, for the size 10 × 2 × 10—SO, SX and AX-s perform better than the rest of the algorithms.

Table 11.7 The algorithms ordered non-decreasingly according to their RE_{min}, RE_{avg} and RE_{max} for the size 10 × 2 × 50 of the problem Θ_z

Nr	Alg-m	RE_{min} (%)	Alg-m	RE_{avg} (%)	Alg-m	RE_{max} (%)
1	TS	0,00	AX-m	2,47	SO	8,69
2	SX	0,00	AX-s	2,53	TS	8,79
3	AX-m	0,00	SO	2,77	SX	9,23
4	AX-s	0,00	SX	2,79	AX-m	11,02
5	SO	0,00	TS	3,09	AX-s	11,19
6	AO-m	0,00	AO-m	5,77	AO-s	15,41
7	AO-s	0,03	AO-s	5,78	AO-m	16,61

Table 11.8 The algorithms ordered non-decreasingly according to their RE_{min}, RE_{avg} and RE_{max} for the size 10 × 3 × 50 of the problem Θ_z

Nr	Alg-m	RE_{min} (%)	Alg-m	RE_{avg} (%)	Alg-m	RE_{max} (%)
1	AX-s	0,00	AX-m	3,31	AX-m	14,66
2	SX	0,00	AX-s	3,46	AX-s	16,59
3	AO-s	0,00	SO	3,86	SX	17,34
4	AX-m	0,00	SX	4,18	TS	18,16
5	SO	0,00	TS	5,72	AO-s	25,24
6	AO-m	0,00	AO-s	5,86	AO-m	27,02
7	TS	0,06	AO-m	5,91	SO	36,35

Table 11.9 The algorithms ordered non-decreasingly according to their RE_{min}, RE_{avg} and RE_{max} for the size 20 × 2 × 50 of the problem Θ_z

Nr	Alg-m	RE_{min} (%)	Alg-m	RE_{avg} (%)	Alg-m	RE_{max} (%)
1	SO	0,00	AX-m	3,84	SX	11,31
2	SX	0,00	AX-s	3,95	SO	11,77
3	AX-m	0,00	SO	5,18	AX-m	12,44
4	AX-s	0,00	SX	5,33	AX-s	12,49
5	AO-m	0,87	TS	6,19	TS	12,65
6	AO-s	0,96	AO-s	7,73	AO-s	17,45
7	TS	1,05	AO-mm	7,80	AO-m	18,76

For the problem size 10 × 3 × 10, according to the values of the RE_{min} in Table 11.2, algorithms AX-s, SO, AX-m were able to find the best-known solutions, and algorithms: SX, TS, AO-m, AO-s could not. However, RE_{min} values of the latter group do not differ significantly from the best-known solutions, namely, from 0,01% to 0,28%. Considering RE_{avg}, SO algorithm has the lowest RE_{avg}, and this way, is the leader in a group of algorithms: SO, AX-m, SX, AX-s with RE_{avg} values differing from 4,67% to 4,74%. TS algorithm is in-between the first group and the third, made of AO-m and AO-s algorithms, whose RE_{avg} values are considerably higher than in the first group, i.e. 8,66% and 8,71% respectively. Speaking about RE_{max}, it is also possible to classify the algorithms into two groups of similar values of RE_{max}: SO, AX-m, TS, SX, from 15,92% to 18,39%, and a group of high RE_{max}: AX-s, AO-s, AO-m, from 19,33% to 26,06%. For the problem size $10 \times 3 \times 10$, we also give the intervals to which belong the values of the considered parameters, i.e. $RE_{min} \in [0{,}00\%, 0{,}28\%]$, $RE_{avg} \in [4{,}67\%, 8{,}71\%]$, and finally, $RE_{max} \in [15{,}92\%, 26{,}06\%]$.

As it could be seen, the RE_{avg} and RE_{max} of all algorithms have increased while scheduling 10 tasks on 3 machines compared to scheduling 10 tasks on 2 machines. In this case again, homogeneous ring topology, as well as random topology for both homogeneous and heterogeneous structures performed better than other algorithms. In addition, it can be noticed, that the algorithms where CE procedure sends multiple solutions during the migration phase perform better, than when it sends a single solution. To finalize, for the size $10 \times 3 \times 10$—SO, AX-m and SX perform better, than the other algorithms.

For the problem size $20 \times 2 \times 10$, according to the values of the RE_{min} in Table 11.3, only algorithm AX-m was able to find the best-known solutions, and the rest of the algorithms—could not. According to RE_{min}, the algorithms: AX-s, SX, SO, TS make a middle group with the values from 0,26% to 0,68%. The algorithms AO-m and AO-s make the third group with RE_{min} values from 1,12% to 1,16%, which are nearly twice as high as in the middle group.

Considering RE_{avg}, AX-m algorithm has the lowest RE_{avg}, and is the leader in a group of the algorithms: AX-m, AX-s, SO, SX, TS with RE_{avg} values differing from 4,76% to 6,26%. The algorithms AO-s, AO-m make the third group of the similar RE_{avg} values, where RE_{avg} is from 9,19% to 9,31% which are considerably higher than in the middle group. Considering RE_{max}, it is also possible to classify the algorithms into two groups: with low RE_{max}: SO, SX, AX-m, TS, AX-s, with the values from 11,47% to 12,54%, and another group: AO-m, AO-s, with high RE_{max} values from 16,71% to 18,19%. For the problem size $20 \times 2 \times 10$, the intervals of the REs are as follow: $RE_{min} \in [0{,}00\%, 1{,}16\%]$, $RE_{avg} \in [4{,}76\%, 9{,}31\%]$, and finally $RE_{max} \in [11{,}47\%, 18{,}19\%]$. For this problem size, our observations on the topology point at the random topology of both heterogeneous and homogeneous structures, as well as homogeneous ring topology as most efficient ones. Here, CE procedure sending multiple solutions during the migration phase, performs better, than when it sends only a single solution. For the size $20 \times 2 \times 10$, AX-m, SO and AX-s are the most efficient algorithms.

For the problem size $10 \times 2 \times 20$, according to the values of the RE_{min} in Table 11.4, all algorithms were able to find the best-known solutions. Here, $RE_{min} = 0{,}00\%$, $RE_{avg} \in [2{,}05\%, 5{,}01\%]$, $RE_{max} \in [6{,}67\%, 17{,}60\%]$. The algorithms implementing ring or random topologies realized on homogeneous islands, as well as random topology realized on all islands have considerably lower RE_{avg} and RE_{max} in comparison with the algorithms exploiting the ring topology built on all islands. Thus, SO, SX and AX-m algorithms outperform the other algorithms while solving the problem of the size $10 \times 2 \times 20$.

For the problem size $10 \times 3 \times 20$, according to the values of the RE_{min} in Table 11.5, all algorithms, except for AO-s, were able to find the best-known solutions. Here, the values of $RE_{min} \in [0,00\%, 0,08\%]$, $RE_{avg} \in [3,61\%, 8,13\%]$, and $RE_{max} \in [14,71\%, 26,03\%]$. The overall results for the size $10 \times 3 \times 20$ are nearly the same as for $10 \times 2 \times 20$, i.e. the algorithms implementing random or ring topologies realized on homogeneous islands, as well as random topology realized on all islands have considerably lower RE_{avg} and RE_{max} in comparison with the algorithms exploiting the ring topology built on all islands. However, for this size of the problem AX-m algorithm has lower RE_{avg} and RE_{max} than SO. Thus, AX-m, SO and AX-s algorithms outperform the other algorithms while solving the problem of the size $10 \times 3 \times 20$.

For the problem size $20 \times 2 \times 20$, according to the values of the RE_{min} in Table 11.6, all island-based algorithms were able to find the best-known solutions. Here, $RE_{min} \in [0,00\%, 0,53\%]$, $RE_{avg} \in [2,05\%, 5,09\%]$, and finally $RE_{max} \in [9,08\%, 15,61\%]$. The overall results for the size $20 \times 2 \times 20$ are much alike as for $10 \times 2 \times 20$, i.e. the algorithms implementing random or ring topologies realized on homogeneous islands, as well as random topology realized on all islands have considerably lower RE_{avg} and RE_{max} in comparison with the algorithms exploiting the ring topology built on all islands. Again, AX-m algorithm has lower RE_{avg} and RE_{max} than SO for this size of the problem. Thus, AX-m, SO and AX-s algorithms outperform the other algorithms while solving the problem of the size $20 \times 2 \times 20$.

For the problem size $10 \times 2 \times 50$, according to the values of the RE_{min} in Table 11.7, all island-based algorithms, except for AO-s, were able to find the best-known solutions. Here, $RE_{min} \in [0,00\%, 0,03\%]$, $RE_{avg} \in [2,47\%, 5,78\%]$, and finally $RE_{max} \in [8,69\%, 16,61\%]$. The overall results for the size $10 \times 2 \times 50$ show that the algorithms implementing random or ring topologies realized on homogeneous islands, as well as random topology realized on all islands have considerably lower RE_{avg} and RE_{max} in comparison with the algorithms exploiting the ring topology built on all islands. For the size $10 \times 2 \times 50$, the results do not allow to unequivocally determine the most efficient algorithm, thus we distinguish AX-m, SO, SX and AX-s algorithms as more efficient than AO-m and AO-s algorithms.

For the problem size $10 \times 3 \times 50$, according to the values of the RE_{min} in Table 11.8, all algorithms were able to find the best-known solutions. Here, $RE_{min} \in [0,00\%, 0,06\%]$, $RE_{avg} \in [3,31\%, 5,91\%]$, $RE_{max} \in [14,66\%, 36,35\%]$. The overall results for the size $10 \times 3 \times 50$ show that the algorithms implementing random or ring topologies realized on homogeneous islands, as well as random topology realized on all islands have considerably lower RE_{avg} and RE_{max} in comparison with the algorithms exploiting the ring topology built on all islands. For this size of the problem, AX-m, AX-s and SX algorithms outperform the other algorithms while solving the problem.

For the problem size $20 \times 2 \times 50$, according to the values of the RE_{min} in Table 11.9, all algorithms, except for AO-m and AO-s, were able to find the best-known solutions. Here, we give the intervals of the REs' values: $RE_{min} \in [0,00\%, 0,96\%]$, $RE_{avg} \in [3,84\%, 7,80\%]$, $RE_{max} \in [11,31\%, 18,76\%]$. The overall results for the size $20 \times 2 \times 50$ show that the algorithms implementing random or ring topologies realized on homogeneous islands, as well as random topology realized on all islands have considerably lower RE_{avg} and RE_{max} in comparison with the algorithms exploiting the ring topology built on all islands. For the size 20 × 2 × 50, the results do not allow to unequivocally determine the most efficient algorithm, thus we distinguish AX-m, SO, SX and AX-s algorithms as more efficient than AO-m and AO-s algorithms.

In order to determine the percentages of the best found solutions for a particular problem size that are of the same quality as the best-known solutions, we had determined the best solutions found within 43 runs of the algorithm for each of 100 problem instances. Next, we counted how many solutions out of obtained 100 had the same quality as the best known for the same problem size and gave this number in percent. The percentages of the best solutions (PBFS) found by the proposed algorithms, which were of the same quality as the best-known solutions, are given in Tables 11.10, 11.11, and 11.12. The results in Tables 11.10, 11.11, and 11.12 confirm unequivocally the previous conclusion, that the algorithms implementing random or ring topologies realized on homogeneous islands, as well as random topology realized on all islands are more efficient that the algorithms exploiting the ring topology built on all islands. Similarly as for REs, AX-m, AX-s and SX prevail other algorithms with clear dominance of AX-m and AX-s, i.e. the algorithms that implement the random topology realized on all islands.

Table 11.10 The percentage of the best solutions (PBFS) found by the proposed algorithms which have the same quality as the best-known solutions for the discretisation level $W = 10$, ordered non-increasingly

10 × 2 × 10	PBFS (%)	10 × 3 × 10	PBFS (%)	20 × 2 ×10	PBFS (%)
SO	69	AX-m	52	SX	47
SX	68	AX-s	49	SO	28
AX-m	66	SO	46	AX-m	11
AX-s	50	SX	33	AX-s	10
AO-m	44	TS	22	TS	2
AO-s	16	AO-s	8	AO-s	2
TS	8	AO-m	3	AO-m	0

Table 11.11 The percentage of the best solutions (PBFS) found by the proposed algorithms that have the same quality as the best-known solutions for the discretisation level $W = 20$, ordered non-increasingly

10 × 2 × 20	PBFS (%)	10 × 3 × 20	PBFS (%)	20 × 2 × 20	PBFS (%)
AX-m	36	AX-m	53	SX	32
AX-s	36	AX-s	37	AX-m	32
SX	35	SX	27	AX-s	25
SO	31	SO	17	SO	12
TS	16	AO-m	9	AO-m	4
AO-m	10	AO-s	9	AO-s	2
AO-s	8	TS	6	TS	0

Table 11.12 The percentage of the best solutions (PBFS) found by the proposed algorithms that have the same quality as the best-known solutions for the discretisation level $W = 50$, ordered non-increasingly

10 × 2 × 50	PBFS (%)	10 × 3 × 50	PBFS (%)	20 × 2 × 50	PBFS (%)
AX-m	42	AX-m	37	AX-m	47
AX-s	30	AX-s	27	AX-s	40
SO	18	SX	17	SX	9
SX	18	AO-s	12	SO	3
TS	4	SO	10	TS	1
AO-m	3	AO-m	6	AO-m	0
AO-s	1	TS	0	AO-s	0

Although, it was possible to determine several most efficient algorithms for each conducted test, we still can't distinguish the most efficient one. In order to do so, we need some universal measure which could be applied for evaluation of the proposed algorithms. For this reason, we need to transform, or more precisely—normalize RE_{min}, RE_{avg}, RE_{max} and PBFS in a such way, that it would be possible to obtain some estimates that could be aggregated into one estimate, this way enabling the choice of the most efficient algorithm. Because RE and PBFS have opposite evaluation meaning, i.e. the lower RE - the better performance of the algorithm, the lower PBFS - the worse performance of the algorithm, we introduce a new parameter NB = 1 – PBFS instead of PBFS.

Thus, let (11.1) be the formula which we apply to RE_{min}, RE_{avg}, RE_{max} and NB within a particular problem size in order to obtain a normalized estimate *ne*.

$$ne_p = \frac{x - x_{min}}{x_{max}} \tag{11.1}$$

In Eq. (11.1), p—one of the considered parameters, i.e. RE_{min}, RE_{avg}, RE_{max} or NB, x—the value of the considered parameter of the particular algorithm, x_{min}, x_{max}—the minimum or respectively maximum value of the considered parameter within the same problem size among all considered algorithms.

After calculating *ne* values for all parameters of all algorithms for all problem sizes, the values obtained for each algorithm were summed into an aggregated estimate. The values of the aggregated estimates were used to make a ranking of the considered algorithms, which is shown in Table 11.13. As it could be seen in Table 11.13, the ranking implies the superiority of the algorithms implementing random topology realized on both heterogeneous and homogeneous islands over the algorithms implementing the ring topology. Thus, according to the ranking AX-m algorithm is the most efficient among all considered algorithms.

As it could be seen from the experimental results described above, it is possible to reduce the REs of the solutions found by the considered algorithms just by changing the interconnection topology of the constituent islands. Tables 11.14 and 11.15 show that by changing the interconnection topology RE_{avg} can be reduced by 2,59–4,64% and RE_{max} by 6,53–21,69% dependently on the problem size. The

Table 11.13 A ranking of the considered algorithms according to the aggregated estimate values

Algorithm	Aggregated estimate	Ranking
AX-m	0,83	1
AX-s	2,05	2
SX	3,06	3
SO	4,94	4
TS	9,61	5
AO-m	13,68	6
AO-s	16,50	7

Table 11.14 The ranges and deltas of RE_{avg} values for the considered problem sizes ordered by ΔRE_{avg} non-decreasingly

Problem size	RE_{avg} range (%)	ΔRE_{avg} (%)
20 × 2 × 20	2,05–5,09	2,59
10 × 3 × 50	3,31–5,91	2,60
10 × 2 × 20	2,05–5,01	2,96
10 × 2 × 10	3,28–6,30	3,02
10 × 2 × 50	2,47–5,78	3,31
20 × 2 × 50	3,84–7,80	3,96
10 × 3 × 10	4,67–8,71	4,04
10 × 3 × 20	3,61–8,13	4,52
20 × 2 × 10	4,76–9,31	4,64

Table 11.15 The ranges and deltas of RE_{max} values for the considered problem sizes ordered by ΔRE_{max} non-decreasingly

Problem size	RE_{max} range (%)	ΔRE_{max} (%)
20 × 2 × 20	9,08–15,61	6,53
20 × 2 × 10	11,47–18,19	6,72
10 × 2 × 10	9,76–16,53	6,77
20 × 2 × 50	11,31–18,76	7,45
10 × 2 × 50	8,69–16,61	7,92
10 × 3 × 10	15,92–26,06	10,14
10 × 2 × 20	6,67–17,60	10,93
10 × 3 × 20	14,71–26,03	11,32
10 × 3 × 50	14,66–36,35	21,69

RE_{avg} and RE_{max} ranges were taken from Tables 11.1, 11.2, 11.3, 11.4, 11.5, 11.6, 11.7, 11.8, and 11.9. Similarly, Table 11.16 shows that the percentage of the best found solutions that have the same quality as the best-known solutions can be increased by 28–61% dependently on the problem size. The PBFS ranges were taken from Tables 11.10, 11.11, and 11.12.

Finally, in Tables 11.17 and 11.18, we observe the influence of the level of the continuous resource discretisation W on the REs of the found solutions. In Table 11.17, for the problem size $10 \times 2 \times W$, $W \in \{10, 20, 50\}$, for almost all algorithms except for AO-s, both RE_{min} and RE_{avg} have the lowest values when $W = 20$. Thus, the influence of W on the REs for the considered problem size could be generalized by the following relations:

Table 11.16 The range and delta of PBFS values for the considered problem sizes ordered by ΔPBFS non-decreasingly

Problem size	PBFS range (%)	ΔPBFS (%)
10 × 2 × 20	8–36	28
20 × 2 × 20	2–32	30
10 × 3 × 50	6–37	31
10 × 2 × 50	1–42	41
10 × 3 × 20	9–53	44
20 × 2 × 10	0–47	47
20 × 2 × 50	0–47	47
10 × 3 × 10	3–52	49
10 × 2 × 10	8–69	61

$$RE_{min/avg}(W=20) < RE_{min/avg}(W=50) < RE_{min/avg}(W=10) \quad (11.2)$$

For the RE_{max}, the results are mixed and it's impossible to derive one clear rule for all algorithms. The influence of the discretisation level W on the REs of the solutions found by the considered algorithms for the problem size 10 × 3 × W, $W \in \{10, 20, 50\}$, according to Table 11.18 could be generalized by the following relations:

$$RE_{min}(W=20) \leq RE_{min}(W=50) < RE_{min}(W=10), \text{ except for AO-s} \quad (11.3)$$

and

$$RE_{avg}(W=50) < RE_{avg}(W=20) < RE_{avg}(W=10), \text{ except for SX and TS} \quad (11.4)$$

For RE_{max}, the results are mixed and it's impossible to derive one clear rule for all algorithms.

The influence of the discretisation level W on the REs of the solutions found by the considered algorithms for the problem size 20 × 2 × W, $W \in \{10, 20, 50\}$, according to Table 11.19 could be generalized by the following relations:

$$RE_{min}(W=20) \leq RE_{min}(W=50) < RE_{min}(W=10) \text{ except for TS} \quad (11.5)$$

and

$$RE_{avg}(W=20) < RE_{avg}(W=50) < RE_{avg}(W=10) \quad (11.6)$$

For RE_{max}, the results are mixed and it's impossible to derive one clear rule for all algorithms. Below we tabularize the obtained relations together in Table 11.20.

As it could be seen in Table 11.20, it's impossible to determine unequivocally the discretisation level on which REs of the found solutions are the lowest. However, it follows from the obtained results, that the relation REs (W = 20) < REs (W = 50) < REs (W = 10) is the most frequent one. This might impose the

Table 11.17 The influence of the level of the continuous resource discretisation W on the REs of the solutions for the problem size $10 \times 2 \times W$, $W \in \{10, 20, 50\}$

Alg-m	RE_{min}			RE_{avg}			RE_{max}		
	$W = 10$ (%)	$W = 20$ (%)	$W = 50$ (%)	$W = 10$ (%)	$W = 20$ (%)	$W = 50$ (%)	$W = 10$ (%)	$W = 20$ (%)	$W = 50$ (%)
AO-m	0,01	0,00	0,00	6,30	4,99	5,77	14,68	17,18	16,61
AO-s	0,01	0,00	0,03	6,30	5,01	5,78	16,53	17,60	15,41
AX-m	0,01	0,00	0,00	3,54	2,27	2,47	12,33	9,14	11,02
AX-s	0,01	0,00	0,00	3,58	2,33	2,53	11,39	12,40	11,19
SO	0,19	0,00	0,00	3,28	2,05	2,77	9,76	6,67	8,69
SX	0,01	0,00	0,00	3,31	2,07	2,79	10,00	7,04	9,23
TS	0,01	0,00	0,00	3,49	2,36	3,09	9,91	7,66	8,79

Table 11.18 The influence of the level of the continuous resource discretisation W on the REs of the found solutions for the problem size $10 \times 3 \times W$, $W \in \{10, 20, 50\}$

Alg-m	RE_{min}			RE_{avg}			RE_{max}		
	$W = 10$ (%)	$W = 20$ (%)	$W = 50$ (%)	$W = 10$ (%)	$W = 20$ (%)	$W = 50$ (%)	$W = 10$ (%)	$W = 20$ (%)	$W = 50$ (%)
AO-m	0,10	0,00	0,00	8,66	8,13	5,91	26,06	24,24	27,02
AO-s	0,28	0,08	0,00	8,71	8,13	5,86	25,67	26,03	25,24
AX-m	0,00	0,00	0,00	4,68	3,61	3,31	16,07	14,71	14,66
AX-s	0,00	0,00	0,00	4,74	3,73	3,46	19,33	15,89	16,59
SO	0,00	0,00	0,00	4,67	3,87	3,86	15,92	15,18	36,35
SX	0,01	0,00	0,00	4,69	4,06	4,18	18,39	16,66	17,34
TS	0,07	0,00	0,06	5,36	4,95	5,72	17,72	18,28	18,16

Table 11.19 The influence of the level of the continuous resource discretisation W on the REs of the found solutions for the problem size $20 \times 2 \times W$, $W \in \{10, 20, 50\}$

Alg-m	RE_{min}			RE_{avg}			RE_{max}		
	$W = 10$ (%)	$W = 20$ (%)	$W = 50$ (%)	$W = 10$ (%)	$W = 20$ (%)	$W = 50$ (%)	$W = 10$ (%)	$W = 20$ (%)	$W = 50$ (%)
AO-m	1,12	0,00	0,87	9,31	2,60	7,80	16,71	15,19	18,76
AO-s	1,16	0,00	0,96	9,19	2,72	7,73	18,19	15,61	17,45
AX-m	0,00	0,00	0,00	4,76	2,05	3,84	11,81	9,08	12,44
AX-s	0,26	0,00	0,00	4,86	2,53	3,95	12,54	9,90	12,49
SO	0,51	0,00	0,00	5,46	2,43	5,18	11,47	9,76	11,77
SX	0,40	0,00	0,00	5,56	2,84	5,33	11,74	11,00	11,31
TS	0,68	0,53	1,05	6,26	5,09	6,19	12,09	10,81	12,65

Table 11.20 The relations among the REs of different discretisation levels W, $W \in \{10, 20, 50\}$, for the considered problem sizes

Problem size	Relations among the REs of different discretisation levels W, $W \in \{10, 20, 50\}$
$10 \times 2 \times W$	$RE_{min}(W = 20) < RE_{min}(W = 50) < RE_{min}(W = 10)$ $RE_{avg}(W = 20) < RE_{avg}(W = 50) < RE_{avg}(W = 10)$ RE_{max} – mixed
$10 \times 3 \times W$	$RE_{min}(W = 20) \leq RE_{min}(W = 50) < RE_{min}(W = 10)$, except for AO-s $RE_{avg}(W = 50) < RE_{avg}(W = 20) < RE_{avg}(W = 10)$, except for SX and TS RE_{max} – mixed
$20 \times 2 \times W$	$RE_{min}(W = 20) \leq RE_{min}(W = 50) < RE_{min}(W = 10)$ except for TS $RE_{avg}(W = 20) < RE_{avg}(W = 50) < RE_{avg}(W = 10)$ RE_{max} – mixed

conclusion, that high discretisation level does not ensure the lowest values of the REs and the additional research is needed to identify the most appropriate discretisation of the continuous resource.

11.2.2 Conclusions from the Experiment

The main goal of our research was to find out whether the topology of learning stages (or islands), might have some effect on the algorithm's efficiency. For this reason we proposed six versions of the PLA2 that differ from each other by their structure and migration scheme. The most important conclusion that can be drawn from the experimental results is that the interconnection topology of the constituent islands might have a noticeable impact on the quality of the solutions yielded by the PLA2. It is possible to reduce the relative errors of the solutions found by the PLA2: RE_{avg} by 2,59–4,64% and RE_{max} by 6,53–21,69% dependently on the problem size. Similarly, the percentage of the best found solutions that have the same quality as the best-known solutions can be increased dependently on the problem size by 28–61%. The ranking of the considered algorithms, which was designed to reveal the most efficient interconnection topology, implies the superiority of the algorithms implementing random topology realized on all available islands, i.e. heterogeneous and homogeneous, or exclusively on homogeneous islands, over the algorithms implementing the ring topology. However, the algorithm implementing the directed ring topology realized exclusively on homogeneous islands for some problem sizes yielded solutions that had the lowest RE_{avg} and RE_{max} values. It should be mentioned here, that all the conclusions are valid for particular implementations of the algorithms used in the experiments. The values of some parameters of the algorithms were determined during their tuning and should be determined on the way of an exhaustive experiment.

11.3 Properties of the Island-Based and Single Population Differential Evolution Algorithms

The properties of the island-based and single population differential evolution algorithms that may have a significant influence on their performance were investigated in [2] and are discussed in this section.

In this section we study a special method of the evolutionary search, namely, differential evolution method. Differential evolution is a stochastic direct search and global optimization method proposed in [6].

We examine the properties of two models of the DE search: the model based on a single population (the implementation of which we denote as DEA), and the model based on multiple populations, known as the island model. We considered two versions of the island model: with migration of individuals between islands ($\mathrm{IBDEA}^{\mathrm{m}}$) and without migration ($\mathrm{IBDEA}^{-\mathrm{m}}$). In the $\mathrm{IBDEA}^{\mathrm{m}}$, the islands periodically exchange among themselves their best solutions with a migration rate *ex*. We investigated how the effectiveness of the models under consideration depends on such parameters as the size of a single population x_{P}, and in the case of the island model, also the number of islands K and the migration rate *ex*.

The main goal of the research was to determine the setting of the parameters which would provide the highest effectiveness of the considered models. The secondary objective was to contribute to the knowledge on the behavior of a single and multiple population algorithms with respect to the parameters under concern.

As a test problem, we used the discrete-continuous scheduling problem with continuous resource discretisation (DCSPwCRD), described in Sect. 9.2.2. In order to conduct our tests, we used the differential evolution algorithm (DEA) and the island-based differential evolution algorithm (IBDEA) for solving the DCSPwCRD described in Sect. 10.5.

11.3.1 Computational Experiment

11.3.1.1 Parameter Set Up

In our experiments the values of the parameters of the DEA were assumed to be the same as in [7], namely the scale factor A which controls the evolution rate of the population was set to $A = 1{,}5$ and values of the variable $rand \in [0, 1]$. The crossover constants Cr_p and Cr_l which control the probability that the trial individual will receive the actual individual's tasks or modes were set $Cr_p = 0{,}2$ and $Cr_l = 0{,}1$, where p and l in the notations Cr_p and Cr_l stand for tasks positions and modes respectively. An initial population of feasible individuals in the DEA was generated using the uniform distribution equal $1/n$ for the tasks, and $1/W$ for the task's modes. Our assumptions concerning the test problem are as follows. We considered three combinations of $n \times m$: 10×2, 10×3, and 20×2, where

n is the number of tasks and m is the number of machines, and three levels of the continuous resource discretisation W: 10, 20, 50. This way we considered nine sizes $n \times m \times W$ of the problem Θ_Z: $10 \times 2 \times 10$, $10 \times 2 \times 20$, $10 \times 2 \times 50$, $10 \times 3 \times 10, \ldots, 20 \times 2 \times 50$. For each of the sizes, we considered 6 instances of the problem Θ_Z, which makes 54 instances of the problem in total. These 54 instances were used for testing one value of the parameter under consideration, where the considered parameters were: x_P—the size of population on an island, K—the number of islands, and ex—the migration rate. The schedule lengths determined by the algorithm for these 54 instances were summed up and the obtained result was used for evaluation and comparison purposes. To make the tests credible, the tested algorithm was always run with the same seed of the random number generator. Thus, the only factor that could cause change in the results was the valuc of the parameter that was tested.

In the experiment, the numbers of the fitness function evaluations $\#ev$ necessary for the DEA and the IBDEA to yield one solution was set to $\#ev = 10^6$.

In our tests, we considered different sizes of population x_P on an island and numbers of islands K. The values of x_P, the ranges of K, and max x_A (where $x_A = K \cdot x_P$ is the size of archipelago), used in the tests are given in Table 11.21.

All tests were carried out on a PC under 64-bit operating system Windows 7 Enterprise with Intel(R) Core(TM) i5-2300 CPU @ 2.80 GHz 3.00 GHz, RAM 4 GB compiled with aid of Borland Turbo Delphi for Win32. When the number of fitness function evaluations was set to 720000, mean time required by the DEA to find a solution for the problem sizes 10×2 and 10×3 for all discretisation levels

Table 11.21 The values of x_P and the ranges of K and x_A used in the tests

x_P	min K	max K	min x_A	max x_A
5	2	2150	10	10750
10	2	1062	20	10620
17	2	565	34	9605
27	2	237	54	6399
44	2	237	88	10428
71	2	126	142	8946
115	2	92	230	10580
186	2	49	372	9114
301	2	36	602	10836
487	2	19	974	9253
788	2	14	1576	11032
1275	2	10	2550	12750
2063	2	5	4126	10315
3338	2	5	6676	16690
5401	2	3	10802	16203
8739	2	3	17478	26217
14140	2	2	28280	28280

was approximately 2–3 s and for the problem size 20 × 2 for all discretisation levels approximately 5–6 s. The total time taken by the DEA to process all 54 instances was approximately 206 s.

11.3.1.2 Test Results

The aim of the experiment was to compare the effectiveness of two models of DE search. We compared DE based on a single population, implemented as a differential evolution algorithm (DEA) and DE based on multiple populations, implemented as an island-based differential evolution algorithm (IBDEA). The island model was implemented in two versions: with and without solution migration among the islands, denoted in the paper as IBDEA^{m} and $\text{IBDEA}^{-\text{m}}$ respectively. In the IBDEA^{m}, the islands exchanged their best solutions with different rates. In our tests, the exchange of best solutions was carried out after 1, 2, 3, 4, 5, 7, 9, 11, 13, 15, 17 populations had been generated on each island. We will be using the notation *ex##* for denoting the migration rate, where ## denotes the number of populations after which the migration had been carried out.

Figures 11.6 and 11.7 present the results of our tests. Figure 11.7 illustrates in more detail the results obtained for test sets 92–137 shown in Fig. 11.6. Our observations on the results of the tests are as follows.

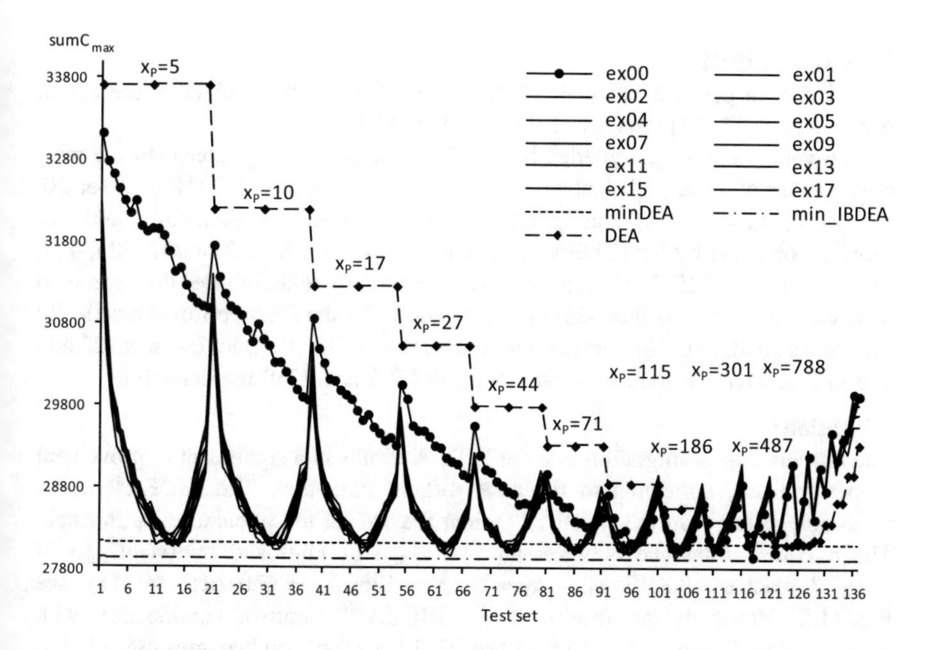

Fig. 11.6 The results obtained for the test sets 1–137

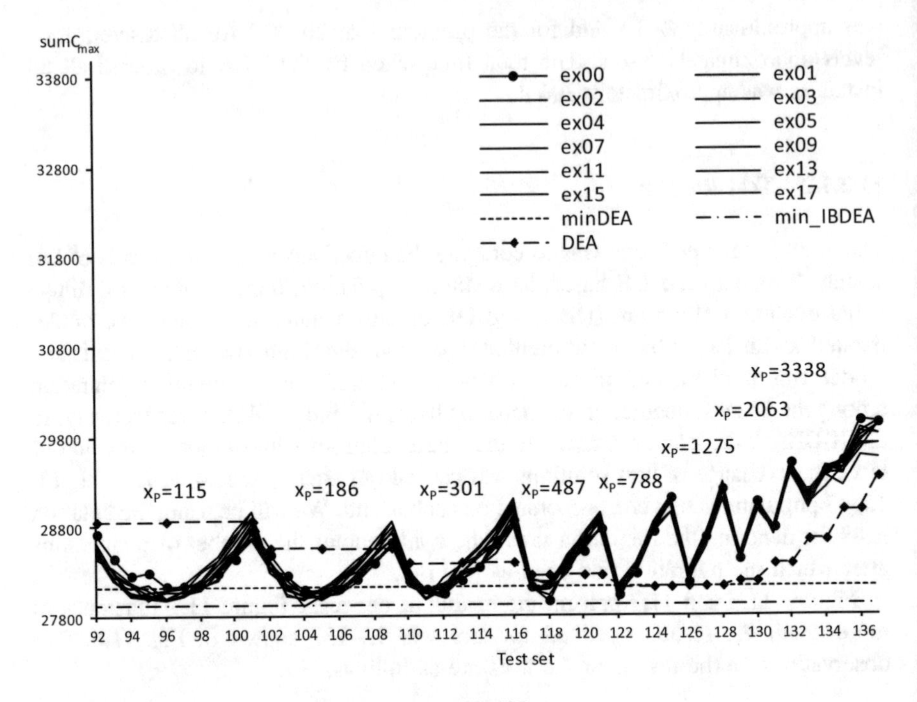

Fig. 11.7 The results obtained for the test sets 92–137

DEA versus IBDEA.
Both IBDEAs perform much better than the DEA when populations are small, namely, $x_P \in [5, 71]$ (test sets 1-91), see Fig. 11.6.

The value of $sumC_{max}$ yielded by the DEA for $x_P = 5$ is greater than the minimum value of $sumC_{max}$ obtained by the IBDEA^{-m} by about 8,8% (test set 20, $K = 2150$, $x_A = 10750$) and about 20% greater than the minimum value of $sumC_{max}$ obtained by the IBDEA^{m} (test sets 13, 14, 15, where $K = 237$, 325, 445, and $x_A = 1185$, 1625, 2225 respectively). However, these differences decrease with increase of x_P. When x_P increases up to 788 and 1275, the DEA performs practically the same as both IBDEAs, see the minimum values of $sumC_{max}$ for test sets 122 and 126 in Fig. 11.7. In these two test sets $x_A = 1576$ and 2250 respectively.

Migration.
The introduction of migration into the IBDEA results in a significant improvement in performance, compared to the case without migration. The IBDEA^{m} works noticeably better than the IBDEA^{-m} when the size of the population x_P is small. The maximal difference between min($sumC_{max}$) of *ex*00 and min($sumC_{max}$) of *ex*01-17 reaches 13,81% when $x_P = 5$, $K = 126$, $x_A = 630$ (test set 11), see Fig. 11.6. However, the results of the IBDEA^{-m} improve significantly with increase of x_P. When $x_P \geq 115$ (test sets 92–137), migration becomes useless. The

results yielded by the IBDEA^{-m} are practically the same as the results yielded by the IBDEA^{m}, see Fig. 11.7.

In the IBDEA^{m}, the rate of migration has no significant influence on the quality of the results. The curves representing $sumC_{max}$ obtained by the IBDEA^{m} with $ex01$ $-ex17$ differ from each other in the area of the minima at most 0,23%, see Fig. 11.6.

Population Size and Number of Islands.
In the case of the IBDEA^{m}, the values of minima of $sumC_{max}$ for all $x_P < 2500$ are almost the same, they differ at most about 0,6%, see Figs. 11.6 and 11.7. These minima are not dependent on any particular number of islands K, but the number of individuals throughout the archipelago x_A, which can be defined approximately as an interval [1200, 2500]. Hence, in order to ensure maximal performance of the IBDEA^{m}, the values of x_P and K should be selected such, that x_A would take values from this range.

For the IBDEA^{-m}, the number of individuals in the population x_P determines the trend and has a major influence on the quality of the results which are improved together with the increase of the population size. The increase of x_P from 5 to 487 resulted in improvement of the best value of $sumC_{max}$ by 10,83%. The results can also be improved by increasing K, e.g. for $x_P = 5$ up to 6,89%. However, this possibility of improvement decreases along with the increase of x_P. The increase of K from 2 to 10 islands for $x_P = 301$ resulted in improvement of the value of $sumC_{max}$ only by 0,5%, see test sets 110–112 in Fig. 11.7.

In the case of the DEA, the quality of the results is also improved along with the increase of x_P. The DEA yields the best results when $x_P \approx [800, 2000]$, see test sets 122–130 in Fig. 11.7. The minimum value of the $sumC_{max}$ yielded by the DEA in this range of x_P differs from the minimum $sumC_{max}$ yielded by both IBDEAs only by 0,69%, see Fig. 11.7.

Finally, it must be said that the exaggerated increase of x_P or K causes deterioration of the quality of results. The increase of x_P over 2500 worsens the results yielded by the DEA and both IBDEAs, see test sets 131−137 in Fig. 11.7. The increase of K, leading to the increase of x_A over 2500, also worsens the quality of results yielded by the IBDEA^{m} for all x_P. However, it is only partially true in the case of the IBDEA^{-m}. The increase of K even to its maximum value when $x_P \in [5, 71]$ improves the quality of results. In these cases x_A might take values up to about 10000. When $x_P > 71$, the IBDEA^{-m} starts to behave in the same way as the IBDEA^{m}.

11.3.2 Conclusions from the Experiment

In the preceding section, we examined properties of the two models of evolutionary search: the model based on a single population (DEA) and the model based on multiple populations, known as the island model. Two versions of the island model have been considered: with migration of individuals between islands (IBDEA^{m}) and without migration (IBDEA^{-m}). In the IBDEA^{m}, the islands periodically exchanged among themselves their best solutions. We investigated how the effectiveness of the models under consideration depends on such parameters as the size of a single population x_P, and in the case of the island model, also the number of islands K and migration rate ex. The main goal of this work was to determine the setting of parameters which would provide the highest efficacy of the considered models. Conclusions of our research are as follows.

The general conclusion is that both models can be equally effective when used with proper parameter settings. Therefore none of them is better than the other. However, it follows from the experiment, that when the population size is small the quality of the results yielded by the IBDEA^{m} and IBDEA^{-m} might be better than the quality of the results yielded by the DEA up to 20% and 8,8% respectively. Such advantage of the island model is caused by a possibility to increase the total number of individuals on the archipelago x_A by increasing the number of the islands K and additionally by the migration of individuals among the islands as it is in the IBDEA^{m}. This advantage deteriorates along with the increase of x_P. It should also be mentioned, that the IBDEA^{m} showed practically the same efficacy for all considered migration rates, so it is hard to choose any of them as the most preferable. The approximate values of the parameters for which the algorithms show their best efficacy are as follows. The DEA is most efficacious when $x_P \approx [800, 2000]$, the IBDEA^{m} and the IBDEA^{-m}—when $x_A \approx [1200, 2500]$, however in the case of the IBDEA^{-m}, it is true when $x_P > 100$. It should be added, that there is no reason of using solution migration mechanism in the island model for the indicated values of x_A. Thus, it would be more practical to use the DEA with $x_P \approx [800, 2000]$, as the simpler implementation of DE.

Finally, we are obliged to emphasize, that the results and conclusions following from them, are true for the number of fitness function evaluations #*ev*, which in our experiments has been set to $\#ev = 10^6$. Properties of the algorithms may differ from those observed in our experiment, when the search process will be limited by smaller values of #*ev*.

11.4 Improving Performance of the Differential Evolution Algorithm Using Cyclic Decloning and Changeable Population Size

The research on performance improvement of the differential evolution algorithm using cyclic decloning and changeable population size has been done in [3] and is discussed in this section.

During search, performed by the evolutionary algorithm (EA), it often comes to the point when individuals in the population reach a configuration such that evolutionary operators no longer produce offspring that can outperform their parents [8]. This phenomenon is known as convergence of EAs. The natural consequence of convergence is that in the population grows the number of similar or identical individuals. An important concern and shortcoming of EAs is premature convergence, i.e. convergence to local optima. EA operating on the population with low diversity sticks in local optima, which impedes searching for global ones. Thus, ensuring high diversity of the population is viewed by researchers as one of the essential factors that affect EAs performance. To prevent getting trapped in local optimum, several approaches have been proposed in the literature. Short review of these approaches can be found in Sect. 9.8.

In this paper we study a special case of the evolutionary algorithm, which is differential evolution algorithm. Differential evolution is a stochastic direct search and global optimization algorithm proposed in [6].

For our experimental research on preserving population diversity, we designed a decloning procedure which is supposed to cyclically replace genetically identical individuals (clones) with randomly generated ones, the detailed description of the decloning procedure is given in Sect. 11.4.2.

The goal of our research was to investigate the extent to which performance of the considered differential evolution algorithm depends on such parameters as the population diversification rate, the size of the population, and the number of fitness function evaluations carried out by the algorithm to yield a solution to the problem.

The goal of the experiments was to determine the most advantageous, in terms of the algorithm's performance, values of mentioned parameters. The results obtained through our experiments allow us to state, that the main contribution of the discussed work is the increase of the range of tools and methods for preserving diversity of the population undergoing DE, as well as proposing the performance improvement policy which takes advantage of experimentally determined relationships between the performance of DE and considered parameters. We also propose a decloning procedure which was used in the experiments for cyclic population diversification. The procedure extends packing technique proposed in [9], ROG technique proposed in [10], and $(\mu + 1)$ EA with genotype diversity proposed in [11]. In the notation $(\mu + \lambda)$, μ stands for the size of the population and λ—for the number of offspring generated at a time from parents.

11.4.1 Computational Experiment

In our experiments on efficiency improvement, values of the parameters of the DEA were assumed to be the same as in [7], namely the scale factor A which controls the evolution rate of the population was set to $A = 1{,}5$ and values of the variable *rand* $\in [0, 1]$. The crossover constants Cr_p and Cr_l which control the probability that the trial individual will receive the actual individual's tasks or modes were set $Cr_p = 0{,}2$ and $Cr_l = 0{,}1$, where p and l in the notations Cr_p and Cr_l stand for tasks positions and modes respectively. In our experiments, we considered the following population sizes: 20, 50, 100, 200, 1000, and the numbers of the fitness function evaluations necessary for the DEA to yield one solution: 37800, 450000, and 720000. An initial population of feasible individuals in the DEA was generated using the uniform distribution equal $1/n$ for the tasks, and $1/W$ for the task's modes.

Our assumptions concerning the test problem are as follows. We considered three combinations of $n \times m$: 10×2, 10×3, and 20×2, where n is the number of tasks and m is the number of machines, and three levels of continuous resource discretisation W: 10, 20, 50. This way we considered nine sizes $n \times m \times W$ of the problem Θ_Z: $10 \times 2 \times 10$, $10 \times 2 \times 20$, $10 \times 2 \times 50$, $10 \times 3 \times 10$, … , $20 \times 2 \times 50$. For each of the sizes, we considered 6 instances of the problem Θ_Z, which makes a test set of 54 instances of the problem in total. This test set of 54 instances was used for testing each parameter configuration under investigation, and the considered parameters were: the decloning period T^d, population size x_P, and the number of fitness function evaluations carried out by the algorithm to yield a solution to the problem *#ev*.

All tests were carried out on a PC under 64-bit operating system Windows 7 Enterprise with Intel(R) Core(TM) i5-2300 CPU @ 2.80 GHz 3.00 GHz, RAM 4 GB compiled with aid of Borland Turbo Delphi for Win32. When *#ev* was set to 720000, mean time required by the DEA to find a solution for the problem sizes 10×2 and 10×3 for all discretisation levels was approximately 2–3 s and for the problem size 20×2 for all discretisation levels approximately 5–6 s. The total time taken by the DEA to process all 54 instances was approximately 206 s.

11.4.2 Decloning Procedure

Our experiments on population diversification we conducted using the decloning procedure (DP) that was invoked in regular cycles with the duration of the period determined as the number of fitness function evaluations. The number of calls of the DP depended on the duration of the decloning period and *#ev* that were allowed for the DEA. We assumed, that the DP would remove 100% of identified clones, and replace them by randomly generated individuals (solutions). In these new solutions, the order of the tasks and task's execution modes were determined at random.

The thorough identification of clones in the population might be time consuming and therefore extend the time needed to execute the algorithm. In order to avoid that, we designed a simpler and quicker decloning procedure that identifies clones in an approximate way. We assumed, that a solution is a clone of another solution, if the following conditions hold:

- the fitness function value of both solutions is the same,
- there exists the same task in both solutions, that is executed in the same mode and that is placed at the same position, chosen at random from the second half of the solution's task list,
- the finish time of the task from the second condition in both solutions is the same.

If at least one of the conditions is not met, then both solutions are different. While the establishment of the first condition is obvious, the establishment of the second and third conditions can be justified by the need to increase the probability of clones identification. Considering tasks only from the second half of the solution in the second condition, on the one hand, simplifies the identification process, on the other hand, used together with the third condition, increases the probability that thc solutions under consideration are clones. It is obvious, that such method does not ensure identification of all clones present in the population. In our experiments, the amounts of clones identified in the same population by the Decloning procedure run multiple times varied within 13% range.

11.4.3 Performance Evaluation Measure

For the purpose of evaluating the effect of the considered parameters on the DEA's performance the parameter $sumC_{max}$ was introduced, which is the total of C_{max} values obtained for the test set of 54 instances of the problem. Thus, in order to carry out a single test for a particular parameter configuration and, therefore, to obtain a single value of $sumC_{max}$, the DEA was run 54 times, each time processing one of 54 test instances. The total value of the obtained 54 schedule lengths created a single value of $sumC_{max}$. To ensure the credibility of results, the tests were always carried out by the DEA with the same seed of the random number generator. Therefore, the only factor that could cause a change in the results was the value of the parameter that was tested. In cases, when the DEA had to be run with the randomized seed of random number generator for evaluation and comparison purposes we used average of the $sumC_{max}$ values (AVG $sumC_{max}$) obtained in multiple tests. The need of such aggregated parameter is justified by the stochastic nature of the DEA, which causes yielding different results for the same input data when the algorithm is run repeatedly. The use of AVG $sumC_{max}$ as a reference value will make our observations and further conclusions more credible and allow for a reliable comparison of the algorithm's performance when run with different parameter settings. An example of such comparison is given in Fig. 11.9, where the values of AVG $sumC_{max}$ are shown for the cases when the algorithm was run with and without decloning with the randomized seed of random number generator. In this example, AVG $sumC_{max}$ values in both cases were calculated as the average of 200 $sumC_{max}$ values obtained in 200 tests. We will discuss the results of our tests using figures illustrating the influence of decloning and population size on the DEA's performance. In our experiment, results generated by the DEA are considered to be better, the smaller are the corresponding values of $sumC_{max}$.

11.4.4 Experiments on Decloning

In order to reveal the influence of decloning on the results obtained by the DEA, we run the algorithm multiple times, each time processing the test set of 54 instances of the problem with different rate of decloning. The period of decloning T^d was defined as the number of fitness function evaluations which were carried out between the successive calls of the Decloning procedure. The largest improvement effect of decloning on the results is observed when it is performed frequently, i.e. when decloning period T^d takes small values, see Figs. 11.8, 11.9, 11.10, 11.11,

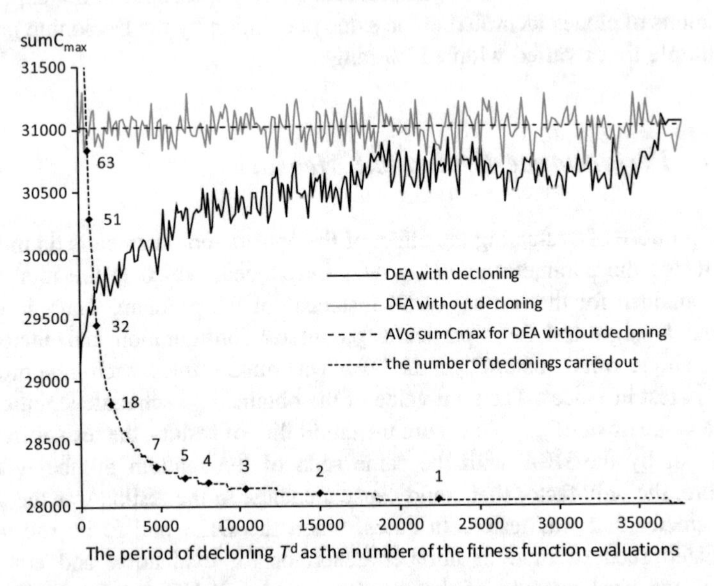

Fig. 11.8 The effect of decloning on $sumC_{max}$, x_P = 20, $\#ev$ = 37800, $T^d \in$ [20, 37800]

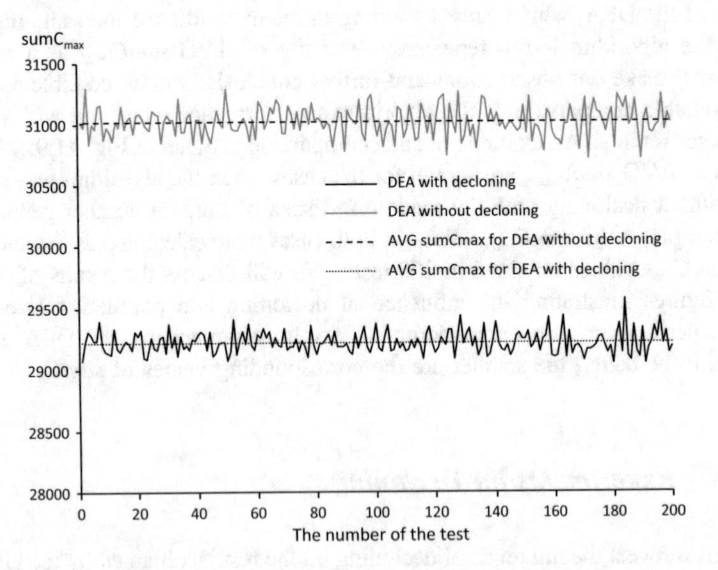

Fig. 11.9 $sumC_{max}$ yielded by the DEA without and with decloning, x_P = 20, $\#ev$ = 37800, T^d = 20

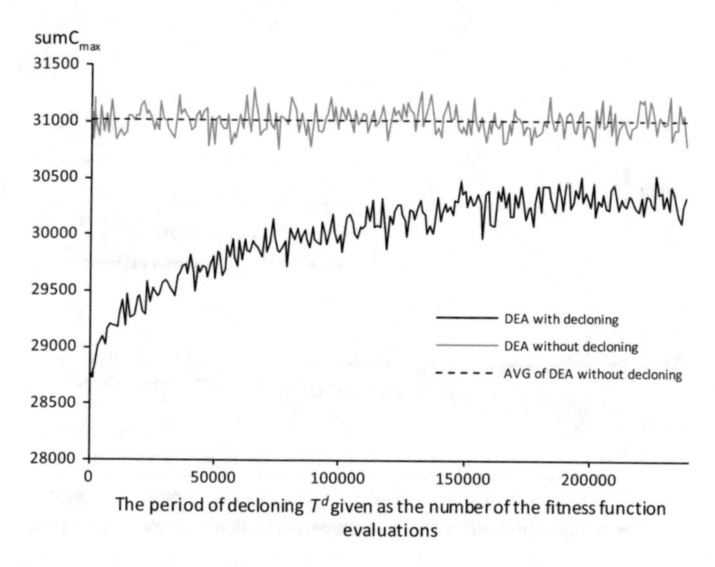

Fig. 11.10 The effect of decloning on $sumC_{max}$, $x_P = 20$, $\#ev = 720000$, $T^{\mathrm{d}} \in [20, 238200]$

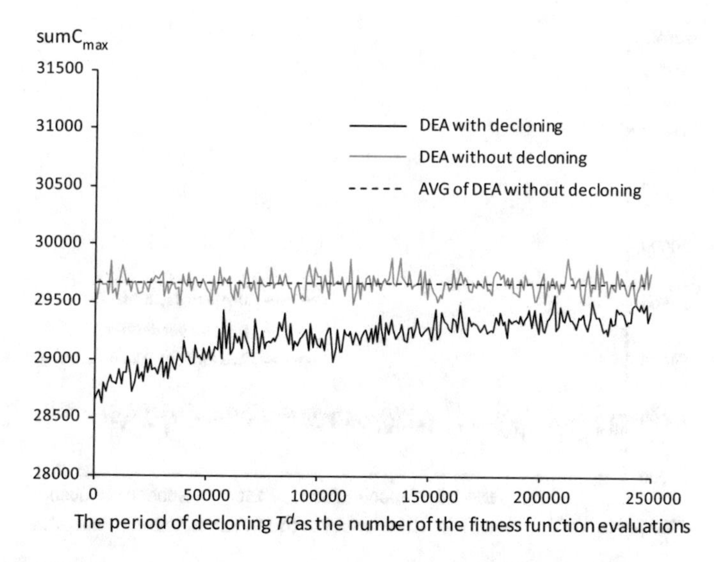

Fig. 11.11 The effect of decloning on $sumC_{max}$, $x_P = 50$, $\#ev = 720000$, $T^{\mathrm{d}} \in [50, 249200]$

11.12, and 11.13. The curves corresponding to the DEA with decloning, were built on the points, each of which is a value of $sumC_{max}$ obtained for different decloning periods. In order to validate the effect of decloning we used AVG $sumC_{max}$ as the reference value, that was determined for presumably the best decloning period, e.g. see Fig. 11.9, in which the lines for the DEA with decloning were obtained for $T^{\mathrm{d}} = 20$. The improvement effect of decloning on the results yielded by the DEA

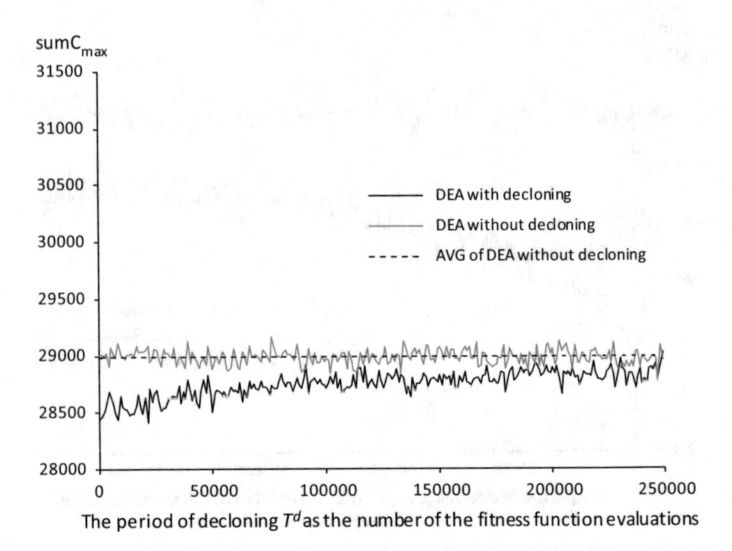

Fig. 11.12 The effect of decloning on $sumC_{max}$, $x_P = 100$, $\#ev = 720000$, $T^d \in [100, 249200]$

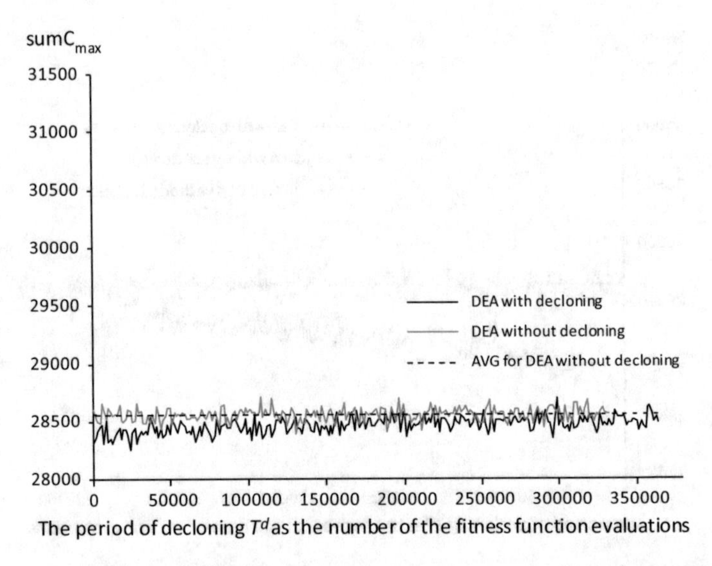

Fig. 11.13 The effect of decloning on $sumC_{max}$, $x_P = 200$, $\#ev = 720000$, $T^d \in [200, 364200]$

given in percent is shown in Fig. 11.15. In Fig. 11.15, we compare values of AVG $sumC_{max}$ determined for all considered population sizes, when the DEA had at its disposal $\#ev = 37800$, $\#ev = 450000$, and $\#ev = 720000$. It follows from Fig. 11.15, that the most significant improvement effect of decloning on the results is observed when the DEA operates on small populations. For example, the curve

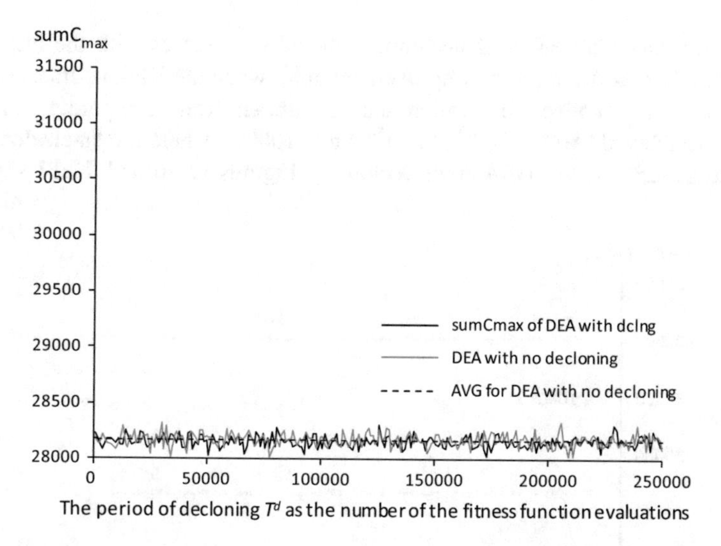

Fig. 11.14 The effect of decloning on $sumC_{max}$, x_P = 1000, $\#ev$ = 720000, $T^d \in$ [1000, 250000]

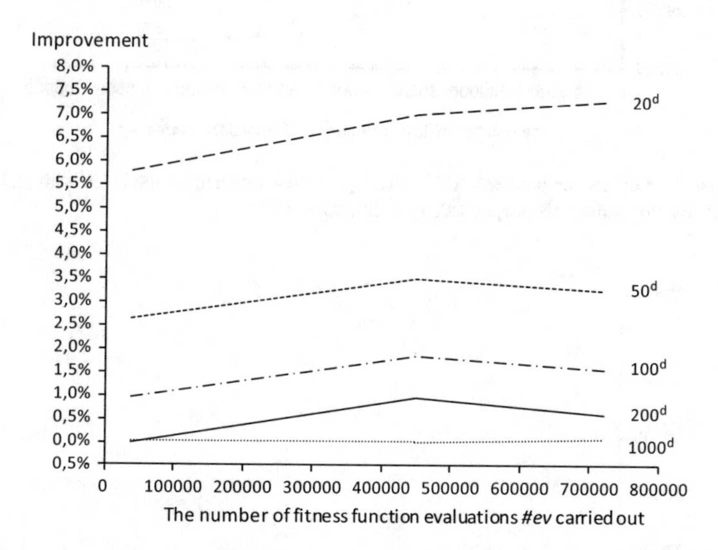

Fig. 11.15 The improvement of the results due to decloning for different population sizes x_P compared to the case without decloning, given in percent

labeled "20^d", which stands for the population size $x_P = 20$ in the DEA with decloning, shows the greatest improvement effect of decloning among all considered population sizes. The results yielded by the DEA with decloning and x_P = 20 were on average 5,79% ($\#ev$ = 37800), 7,01% ($\#ev$ = 450000), and 7,29% ($\#ev$ = 720000) better, than the results yielded without decloning.

The improvement effect of decloning gradually decreases with the increase of population size resulting in no improvement at all when the DEA operates on large population. The above observation can be drawn from comparing curves in Fig. 11.15 labeled "50[d]", "100[d]", "200[d]", and "1000[d]", where the labels denote the population sizes in the DEA with decloning. Figures 11.16 and 11.17 show the

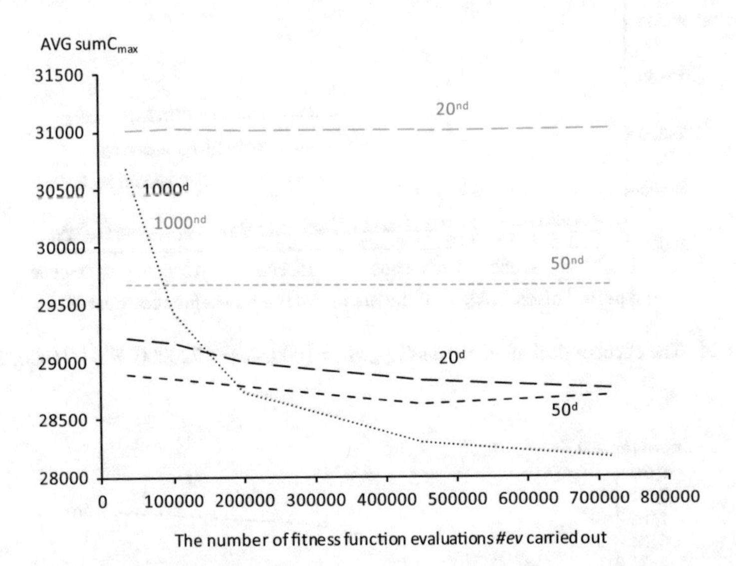

Fig. 11.16 The difference between AVG $sumC_{max}$ values obtained by the DEA with and without decloning for population sizes $x_P = 20$, $x_P = 50$, $x_P = 1000$

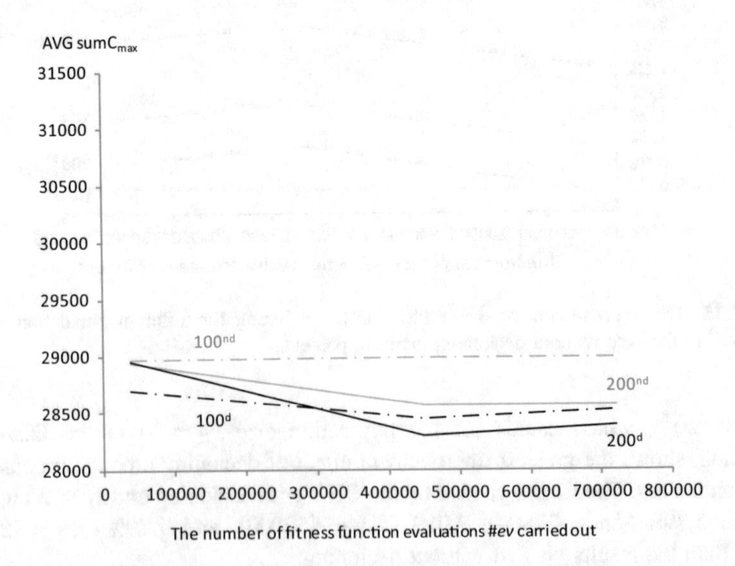

Fig. 11.17 The difference between AVG $sumC_{max}$ values obtained by the DEA with and without decloning for population sizes $x_P = 100$, $x_P = 200$

Table 11.22 Decloning periods T^d, most advantageous for the DEA's performance, given as the number of fitness function evaluations *#ev* and the number of generations in parentheses respectively

#ev	20^d	50^d	100^d	200^d	1000^d
37800	20 (1)	50 (1)	100 (1)	17400 (87)	19000 (19)
450000	420 (21)	50 (1)	100 (1)	200 (1)	21000 (21)
720000	440 (22)	3200 (64)	1000 (10)	2400 (12)	51000 (51)

improvement effect of decloning on AVG $sumC_{max}$, where values AVG $sumC_{max}$ of the DEA with decloning are compared to the values of AVG $sumC_{max}$ yielded by the DEA without decloning (letters "nd" in the upper index of the label denote the cases without decloning). It should be also added, that in all considered cases, except for one, the DEA with decloning performed better than without decloning. The exception is $x_P = 200$ (*#ev* = 37800), where AVG $sumC_{max}$ of the DEA without decloning were better by 0,02%.

The decloning periods T^d, presumably most advantageous for the DEA's performance, determined in our experiments are given in Table 11.22.

11.4.5 Experiments on Population Size

As it has been already observed, increasing the population size weakens the improvement effect of the decloning. Nonetheless, population size is the second important factor contributing to the results improvement. The general observation is that with the growth of population size, the DEA's results become better. When the population size was large, e.g. $x_P = 1000$, see Fig. 11.14, the results obtained were better in comparison to the cases with the smaller population sizes, see Figs. 11.10, 11.11, 11.12, and 11.13. It might seem at this point, that the population size is a principal factor affecting the results improvement, and that there is no need of decloning at all, since it is enough to increase population size to achieve better results. However, this is not always true. It turns out, that there are circumstances in which decloning ensures better results than increasing population size. In order to determine the most beneficial strategy for the results improvement, the third factor, namely *#ev* should be taken into consideration.

11.4.6 Experiments on the Number of Fitness Function Evaluations

Our experiments show, that *#ev* might also contribute to the algorithm's performance improvement. When the DEA without decloning operated on a large population, e.g. $x_P = 1000^{nd}$, the results were improved over 8% merely by the increase

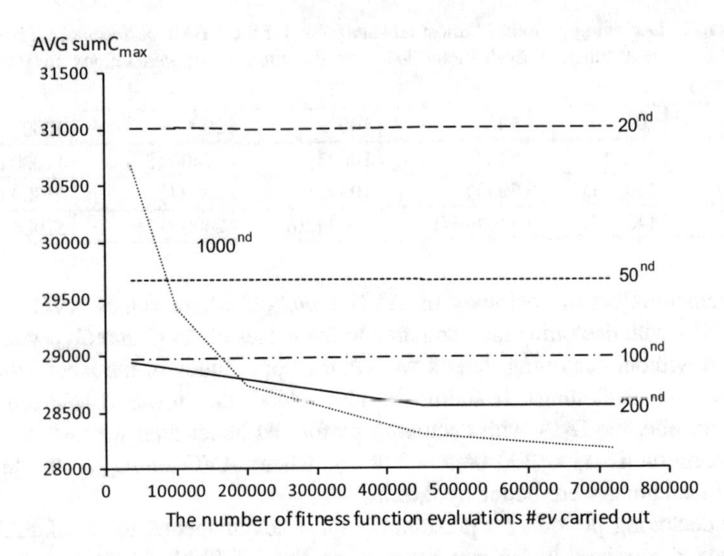

Fig. 11.18 AVG $sumC_{max}$ of the DEA without decloning, considered for different x_P and #ev

of #ev from 37800 to 720000, see Fig. 11.18. The same figure shows the improvement of about 1,3% for $x_P = 200^{nd}$ and no improvement at all for the population sizes $x_P \in \{20^{nd}, 50^{nd}, 100^{nd}\}$. When the DEA operated with decloning, the improving effect, caused by the increase of #ev, was observed for all population sizes, see Fig. 11.19. Although, the greatest improvement effect is observed again for $x_P = 1000^{d}$, we must remind, that for this population size

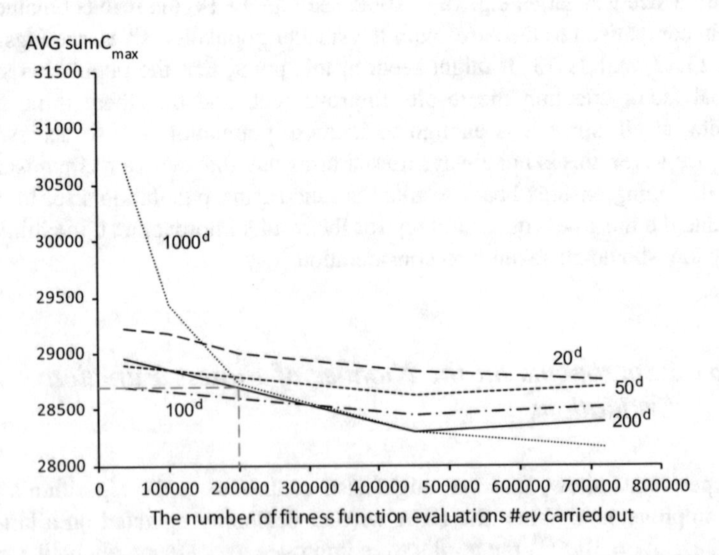

Fig. 11.19 The effect of decloning on AVG $sumC_{max}$, considered for different x_P and #ev

decloning does not contribute to the results improvement, as the curves $x_P = 1000^{nd}$ and $x_P = 1000^{d}$ in Fig. 11.16 are identical.

11.4.7 Performance Improvement Policy

The main goal of our experiments was to find out how to improve the ability of the DEA to yield better solutions. We tried to achieve this goal by making use of decloning and determining most advantageous T^{d}, x_P, and *#ev*. As it follows from the experiment, all three factors, might improve the results, provided that they have been assigned the most advantageous values. Table 11.23 below shows in percent relative difference between the actual AVG $sumC_{max}$, for particular *#ev* and x_P, and the best among all values of AVG $sumC_{max}$ (obtained for *#ev* = 720000 and $x_P = 1000$), provided, that in each considered case, the decloning period T^{d} was assigned the values from Table 11.22.

Thus, the relative difference between AVG $sumC_{max}$ for the case *#ev* = 37800, $x_P = 20$ and the best AVG $sumC_{max}$ is 3,77%. The smallest relative differences for particular *#ev* are given in Table 11.23 in bold font, with the purpose to indicate values of x_P that are most preferable in terms of increasing the performance of the DEA. Therefore, in order to achieve better results, x_P should be determined according to the *#ev* used. It follows from Table 11.23, that if *#ev* increases, then most advantageous values of x_P increase as well.

Based on the above observations, we propose a policy for improving the DEA's performance in terms of the quality of results and, when possible, in terms of response time. In our view the DEA's performance can be improved by adjusting x_P to the available *#ev*, such that will ensure the best results, e.g. if *#ev* is limited to 37800, then $x_P = 100$, if *#ev* = 450000, then $x_P = 200$, and if *#ev* = 720000, then $x_P = 1000$, see Table 11.23.

The response time of the DEA can be improved as follows. Suppose, the DEA is searching for a solution having at its disposal *#ev* = 720000, which, according to Table 11.23, assumes $x_P = 1000$ for the maximal DEA's performance.

Although, $x_P = 1000$ ensures the best results at the end of the computing, however, setting $x_P = 100$ instead of 1000 ensures better AVG $sumC_{max}$ value within *#ev* = 37800 (compare 1,91% versus 8,98% in Table 11.23 and lines 100^{d} and 1000^{d} in Fig. 11.19 respectively). The DEA with $x_P = 1000$ yields the same

Table 11.23 The relative difference between AVG $sumC_{max}$ obtained for particular values of *#ev* and x_P, and the best AVG $sumC_{max}$ obtained for *#ev* = 720000 and $x_P = 1000$

#ev	20^{d} (%)	50^{d}(%)	100^{d} (%)	200^{d} (%)	1000^{d} (%)
37800	3,77	2,62	**1,91**	2,83	8,98
450000	2,39	1,69	1,02	**0,48**	0,50
720000	2,12	1,91	1,32	0,83	**0,00**

AVG $sumC_{max}$ value only after $\#ev \approx 200000$ (observe two auxiliary perpendicular grey dashed lines in Fig. 11.19). Therefore, setting $x_P = 100$ instead of 1000 for the first $\#ev = 37800$ is an opportunity for shortening the response time of the DEA. Thus, after carrying out $\#ev = 37800$, $x_P = 100$ should be changed into $x_P = 1000$, and the rest of the computing would take $\#ev = 720000 - 200000 \approx 520000$. Now, the same AVG $sumC_{max}$ as for $\#ev = 720000$ would be yielded by the DEA carrying out only $\#ev \approx 37800 + 520000 = 557800$, i.e. $720000/557800 \approx 1{,}29$ times faster.

If the DEA had at its disposal only $\#ev = 450000$, the speed-up would be $\approx 450000/287800 = 1{,}56$. At this point, we wish to draw attention to the fact that the difference in the quality of the results obtained after $ev = 720000$ and $ev = 450000$ is only 0,48%, see Table 11.23.

This fact can be seen as an opportunity to shorten the response time of the algorithm. If a loss of 0,48% on the quality of the results can be tolerated, then carrying out only $ev = 450000$ instead of $ev = 720000$, could shorten the response time of the algorithm $720000/450000 = 1{,}6$ times.

If we additionally apply the proposed policy to the case $ev = 450000$, this would shorten the response time even $720000/287800 \approx 2{,}5$ times. However, applying the policy to the case with $\#ev = 720000$ as well, would reduce the speed-up to $557800 / 287800 \approx 1{,}9$ times. It follows from the above discussion, that the final choice of the values of $\#ev$ and, therefore, x_P should allow to meet someone's expectations as to the quality of results and the response time of algorithm.

Thus, to summarize the above discussion, the performance improvement policy would consist of determining the intervals of $\#ev$ and corresponding to them most advantageous x_P such, that the desired quality of the results and response time of the algorithm is assured. The approximate intervals of $\#ev$ and x_P can be determined using the approach described earlier and the results in Table 11.23 and Fig. 11.19. At this stage of research, validity of the proposed policy should be proved experimentally in each case.

11.4.8 Conclusions from the Experiment

In our research, we investigated the extent to which performance of the considered differential evolution algorithm—the DEA depends on such parameters as the population diversification rate, the size of the population, and the number of fitness function evaluations to yield a solution. In the experiments, the most advantageous values of these parameters, in terms of the algorithm's performance, have been determined, and the improvement policy was proposed. Population diversification was carried out cyclically using the proposed decloning procedure. As the test problem, the discrete-continuous scheduling problem with continuous resource discretisation was used.

The obtained results allowed us to propose a performance improvement policy that might noteworthy improve both the efficacy and response time of the DEA. The idea is to choose the diversification rate, the population size and the number of fitness function evaluations to yield a solution using Tables 11.22 and 11.23, given in Sects. 11.4.4 and 11.4.7 respectively. This might ensure the expected quality of the results and the response time of the algorithm.

The results of our experiments show that the diversification of the population can be preserved in an intensive manner, i.e. using dedicated diversifying mechanisms and procedures, e.g. decloning, or extensive one, by increasing the size of the population. The choice of how to preserve the diversification may depend on the restrictions imposed on the population size, response time, and quality of solutions which should be met by a specific algorithm implementation.

Our population diversification technique differs from the one proposed in [9] ince it uses packing catastrophic operator for replacing genetic, not fitness duplicates. Also, unlike ROG technique proposed in [10], it prevents transition of clones to the next generation due to their fitness, which is allowed in ROG, and unlike in ($\mu + 1$) EA with genotype diversity proposed in [11], a new random individual is introduced every time when a clone is identified, which is not carried out in ($\mu + 1$) EA. Finally, decloning, i.e. population diversification, can be carried out less frequently than every generation, which is the case in ROG and ($\mu + 1$) EA.

Our experiments also show that if a compromise between the quality of the results and the response time of the algorithm can be allowed, then it is possible to significantly reduce the response time using the proposed performance improvement policy while only slightly losing on the quality of the results.

For instance, if one can accept a deterioration in the quality of results by 0,48%, then the response time of the algorithm might be reduced, depending on the assumptions made, from 1,29 to 2,5 times, see Sect. 11.4.7.

The main conclusion which results from our research is that the performance of algorithm does not depend on any single factor considered in the paper, but from all of them, combined together, i.e. is a function of several arguments rather than one. Only the selection of appropriate values of these arguments could meet the expectations as to the effectiveness and the response time of the algorithm. An attempt to mathematically describe how the performance depends on the factors examined in the paper might lay foundations for future work.

References

1. Jędrzejowicz, P., Skakovski, A.: Structure versus efficiency of the cross-entropy based population learning algorithm for discrete-continuous scheduling with continuous resource discretisation. In: Czarnowski, I., Jędrzejowicz, P., Kacprzyk, J. (eds.) Studies in Computational Intelligence. Agent-Based Optimization, vol. 456, pp. 77–102 (2013)
2. Jędrzejowicz, P., Skakovski, A.: Properties of the Island-Based and single population differential evolution algorithms applied to discrete-continuous scheduling. In: Czarnowski, I. et al. (eds.) Intelligent Decision Technologies 2016, Proceedings of the 8th KES International Conference on Intelligent Decision Technologies (KES-IDT 2016)—Part I, Smart Innovation, Systems and Technologies, vol. 56, pp. 349–359 (2016)

3. Jędrzejowicz, P., Skakovski, A.: Improving Performance of the Differential Evolution Algorithm Using Cyclic Decloning and Changeable Population Size. In: Nguyen, N.T., Czarnowski, I., Hwang, D. (eds.), Journal of Universal Computer Science (J.UCS), Special Issue—Computational Intelligence Tools for Processing Collective Data (CITPCD 15), vol. 22(6), pp. 874–893 (2016)
4. Jędrzejowicz, P., Skakovski, A.: A cross-entropy based population learning algorithm for discrete-continuous scheduling with continuous resource discretisation. Neurocomputing **73** (4–6), Special Issue: SI, 655–660 (2010)
5. Różycki, R.: Zastosowanie algorytmu genetycznego do rozwiązywania dyskretno-ciągłych problemów szeregowania. Ph.D. diss., Poznań University of Technology, Poland (2000)
6. Storn, R., Price, K.: Differential evolution—a simple and efficient heuristic for global optimization over continuous spaces. J. Glob. Opt. **11**, 341–359 (1997)
7. Damak, N., Jarboui, B., Siarry, P., Loukil, T.: Differential evolution for solving multi-mode resource-constrained project scheduling problems. Comput. Oper. Res. **36**(9), 2653–2659 (2009)
8. Fogel, D.B.: An introduction to simulated evolutionary optimization. IEEE Trans. Neural Netw. **5**(1), 3–14 (1994)
9. Kureichick, V.M., Melikhov, A.N., Miaghick, V.V., Savelev, O.V., Topchy, A.P.: Some new features in the genetic solution of the traveling salesman problem. Proceedings of the ACEDC'96. Plymouth (1996)
10. Rocha, M., Neves, J.: Preventing premature convergence to local optima in genetic algorithms via random offspring generation. LNAI (Lecture Notes in Artificial Intelligence) **1611**, 127–136 (1999)
11. Friedrich, T., Oliveto, P.S., Sudholt, D., Witt, C.: Analysis of diversity-preserving mechanisms for global exploration. Evol. Comput. **17**(4), 455–476 (2009)

Chapter 12
Conclusions

In the presented study, we have considered the discrete-continuous scheduling problem (DCSP) and approaches for solving it. The discrete-continuous scheduling is scheduling of nonpreemptable tasks on identical machines under constraint of a single renewable continuous resource required for processing the task. The time and the rate of processing the task is a function of the amount of the continuous resource allocated to the task. The amount of the continuous resource is not known in advance and may change during the processing of the task. Thus, the task behavior is determined by the processing rate versus resource amount model which adequately describes the temporary nature of the renewable resource and is more natural in the majority of practical situations. The allocation of the continuous resource to the tasks is, in general, computationally intractable and is obtained by solving the appropriately formulated mathematical programming problem. However, in some special cases with concave power processing rate functions, the optimal resource allocation can be found analytically in polynomial time. These functions are also crucially important from the practical point of view.

Our study is a compilation of theoretical and practical knowledge on the DCSP as well as approaches, methods, and tools used to solve it. The problem is discussed from the point of view of the existing knowledge on the problem, which includes the general approach to solving the DCSP, the properties of the optimal solutions and special classes of the task processing rate functions. Since the DCSP is NP-hard, a variety of heuristic and metaheuristic approaches were developed to solve the problem. We provide the state-of-the-art review on the recent theoretical research on the DCSP as well as existing heuristic and metaheuristic approaches to solve the problem. We also provide a survey of the research on the island model and the convergence in evolutionary algorithms which are important factors in the design of metaheuristics.

The most significant parts of the study are metaheuristics, proposed for solving the DCSP, and the research on their properties and efficiency. Since all proposed metaheuristics implement the island model of computation, we have investigated homogeneous and heterogeneous variants of the model as well as the influence of

E. Ratajczak-Ropel and A. Skakovski, *Population-Based Approaches to the Resource-Constrained and Discrete-Continuous Scheduling*,
Studies in Systems, Decision and Control 108, DOI 10.1007/978-3-319-62893-6_12

the topology of the model on the efficiency of evolutionary search. We have also presented the research on the properties and compared the performance of the DE based on the island model and on a single population. We investigated how the effectiveness of these models depends on such parameters as the size of a single population, and in the case of the island model, also the number of the islands and the migration rate. We have also investigated the extent to which the performance of a single population differential evolution algorithm, proposed for solving the DCSP, depends on such parameters as the population diversification rate and the number of fitness function evaluations. The experimentally determined relationships between the performance of DE and considered parameters allowed us to propose the performance improvement policy, based on using changeable population size and decloning procedure, designed for cyclic diversification of the population. The proposed performance improvement policy has a potential to significantly improve the efficiency of the DE algorithm for solving the DCSP.

It is our hope, that this study will contribute to a better insight into the discrete-continuous scheduling and will increase the range of tools for coping with it. We also hope, that our experimental findings on the metaheuristics' efficiency improvement are not limited to our case, but have more universal meaning and will be still valid for other evolutionary algorithms designed for solving other difficult problems. As it could be seen from the study, both, discrete-continuous scheduling and effective evolutionary algorithm design, are still an open issue. Thus, our future work might involve dealing with various practical situations where findings for the DCSP may be applied to resolve them. Another important direction for future work is to develop methods and explore properties which may contribute to EAs' efficiency improvement.

Erratum to: Population-Based Approaches to the Resource-Constrained and Discrete-Continuous Scheduling

Ewa Ratajczak-Ropel and Aleksander Skakovski

Erratum to:
E. Ratajczak-Ropel and A. Skakovski, *Population-Based Approaches to the Resource-Constrained and Discrete-Continuous Scheduling*, Studies in Systems, Decision and Control 108, DOI 10.1007/978-3-319-62893-6

In the original version of the book, the following corrections have to be incorporated:
In Cover page, book title "Population-Based Approaches to the Resource Constrained and Discrete Continuous Scheduling" has to be changed to read as "Population-Based Approaches to the Resource-Constrained and Discrete-Continuous Scheduling".
In Table of Contents, author names for Part I (Chapters 1–6) and Part II (Chapters 7–12) have to be corrected as "Ewa Ratajczak-Ropel" and "Aleksander Skakovski", respectively.
The erratum book has been updated with the changes.

The updated original online version for this book can be found at
DOI 10.1007/978-3-319-62893-6

E. Ratajczak-Ropel and A. Skakovski, *Population-Based Approaches to the Resource-Constrained and Discrete-Continuous Scheduling*,
Studies in Systems, Decision and Control 108, DOI 10.1007/978-3-319-62893-6_13

Printed in the United States
By Bookmasters